Time Blind

Kevin K. Birth

Time Blind

Problems in Perceiving Other Temporalities

Kevin K. Birth
Department of Anthropology
Queens College, CUNY
Flushing, New York, USA

ISBN 978-3-319-34131-6   ISBN 978-3-319-34132-3 (eBook)
DOI 10.1007/978-3-319-34132-3

Library of Congress Control Number: 2016949030

© Kevin K. Birth 2017
This work is subject to copyright. All rights are solely and exclusively licensed by the Publisher, whether the whole or part of the material is concerned, specifically the rights of translation, reprinting, reuse of illustrations, recitation, broadcasting, reproduction on microfilms or in any other physical way, and transmission or information storage and retrieval, electronic adaptation, computer software, or by similar or dissimilar methodology now known or hereafter developed.
The use of general descriptive names, registered names, trademarks, service marks, etc. in this publication does not imply, even in the absence of a specific statement, that such names are exempt from the relevant protective laws and regulations and therefore free for general use.
The publisher, the authors and the editors are safe to assume that the advice and information in this book are believed to be true and accurate at the date of publication. Neither the publisher nor the authors or the editors give a warranty, express or implied, with respect to the material contained herein or for any errors or omissions that may have been made.

Cover illustration: © Kevin K. Birth

Printed on acid-free paper

This Palgrave Macmillan imprint is published by Springer Nature
The registered company is Springer International Publishing AG
The registered company address is: Gewerbestrasse 11, 6330 Cham, Switzerland

Contents

Preface ix

Prelude: The Duplicity of Time 1

1 (Hegemonic) Calibrations in Anthropology 17

2 Evolution's Anticipation of Horology? 33

3 "Hours Don't Make Work": Kairos, Chronos, and the Spirit of Work in Trinidad 47

4 Past Times: The Temporal Structuring of History and Memory 71

5 Tensions of the Times: Homochronism versus Narratives of Postcolonialism 93

6 Thinking Through Homochronic Hegemony Ethnographically 117

References 141

Index 159

List of Figures

Fig. 2.1	Light/dark cycles of mice at 60° N versus laboratory mice	39
Fig. 2.2	Differences in light between a sunny day and a laboratory	40
Fig. 4.1	Number of remembered events by year	81
Fig. 4.2	Number of remembered events by year related to significant historical events	82
Fig. 4.3	Age at time of remembered event	85
Fig. 4.4	Weighted age at time of remembered event	85
Fig. 4.5	Weighted age at time of remembered event related to culturally marked life stage	87

Preface

The title for this book is influenced by several works that use the metaphor of "blindness" to discuss limits on our scholarly perception. For instance, Bernard Stiegler describes "eidetic blinding" (1998, 3) as the consequence of how the technicization of science has led to a loss of perception, meaning, and memory. Ernst Bloch states, "We ... are located in our own blind spot, in the darkness of the lived moment" (Bloch 2000, 200). Vincent Crapanzano argues that one of the purposes of ethnography is to overcome "cultural blindness" (2003, 4). Finally, in a recent book, my colleague, John Collins, has argued that the tendency of social scientists to grant ontological status to representations without due attention to how those representations are constructed leads to analytic blindness (2015, 362). Cultural ideas of time are crucial elements to the construction of representations—they provide what Hallowell would call a basic orientation provided by culture (1955, 92–100). Social scientists are not immune to the culture that shapes them, even as they seek to overcome the conceptual limitations that culture imposes. One could say that social science suffers from time blindness.

One could say that, but at some risk. Time blindness is a term often applied to one of the manifestations of attention deficit hyperactivity disorder (ADHD). In this use of the term "time blindness," clinicians are describing the difficulty in managing time that many people with ADHD experience. Because of this association, a reviewer of this book suggested that maybe I should rethink the title. I did, but in the end, the publisher and I agreed to keep the title *Time Blind*. I imagine the publisher did so because the title seemed more pithy than my proposed alternative, but

I did so because of the parallel between the disability of time blindness and what I seek to discuss in the book. Those who battle time blindness find it difficult to adhere to the very calendar-driven, clock-shaped sensibility of time that blinds many scholars. At the risk of sounding like R.D. Laing's view of psychosis, those who struggle with ADHD-related time blindness possibly have an insight into time that those mired in the cultural baggage of clock and calendar time do not. This is an interesting possibility, but not one that I am qualified to explore; it does, however, reveal a potential limit of not only understanding, but compassion, that can emerge from the extent to which Enlightenment European cultural ideas of time are unquestionably adopted and enforced.

McLuhan writes, "'Rational,' of course, has for the West long meant 'uniform and continuous and sequential'" (1994, 15). My book criticizes this idea of "rational." Such criticism is influenced by the work of many West Indian thinkers—that European-derived concepts are conceptually limiting. The resistance to clock time and uniform rhythms in order to assert Caribbean sensibilities in the work of C.L.R. James, Édouard Glissant, Antonio Benítez-Rojo, Wilson Harris, Sylvia Wynter, and others has long influenced my ethnographic work on Trinidad and my thinking on time. But I want to go beyond a simple resistance against European hegemony in order to demonstrate the artificiality and contingency of European-derived temporalities that, despite their relatively recent cultural origin, have come to be regarded as natural. So while most of this book is about Trinidad, the arguments are against uniform, homogeneous concepts of time. And these arguments are not merely that the Caribbean, or any other postcolonial context, has its own temporalities, but that homochronicity distorts knowledge in general. "Western" scholarship is built on cultural assumptions about time that are exotic. As Adolf Bastian pointed out, "Sollte, wie jede Frage, die der menschlichen Natur aus der Majorität entschieden werden, so würde Europa, den übrigen Continenten gegenüber, nur als Ausnahme erscheinen, um den Durchschnittsmenschen zu finden" (1860, 230) [Maybe the question about the nature of man is to be decided by the majority; in that case Europeans would be the eccentric ones when compared with average man] (translated in Koepping 1983, 54).

The unusual qualities of European temporalities have been naturalized, and, as Carol Greenhouse has argued, have come to distort ethnographic discussions of time (1996). The process of naturalization began early in our lives as learners. Teaching about "telling time" is a component of early childhood education. In most curricula, this is limited to children learning to interpret representations of time that elide the cultural conventions and rich histories of those representations. For instance, the analog clock represents a Ptolemaic presentation of Egyptian divisions of the day, and Babylonian divisions of the hour shaped by a northern hemispheric bias (a Ptolemaic representation based on the southern hemisphere would have the hours arranged counterclockwise).

The simplicity with which many people approach time then results in difficulty in understanding other temporalities. It is well documented that perception shapes the unfamiliar into the familiar and expected. This is why it is likely that neither I nor the copy editors have caught every typographical error in this book. Yet, it also leads museumgoers to look at a French Revolutionary decimal clock with its ten-hour dial and still see nothing unusual, and it leads generations of ethnographers to convert observed rhythms into clock and Gregorian calendar time. It also leads some of my potential students to wonder if there really is enough material about time to justify an entire class on the topic.

It is not this lack of knowledge about time that concerns me, however. Instead, my concern is how cultural ideas enter into scholarship as unrecognized assumptions. We live on a rotating geoid orbiting the Sun. The cycles and rhythms of our planet have profound biological, economic, political, and cultural consequences. The heritage of European timekeeping distorts our ability to understand the consequences of the Earth's cycles. This is not to belittle the achievements of European horology—quite the contrary, for it demonstrates the far reach of that tradition. But there are other horologies and other ways of thinking about time, and the widespread distribution of Europe's ideas is attributable to European colonial hegemony and not to the superiority of its time-reckoning tradition.

As a discipline that spans biological, historical, and social scholarship, anthropology's holism is well positioned to draw connections between the multiple rhythms and cycles that influence human life. It is for this reason that, while this book ranges far and wide to touch on issues of chronobiology, economics, historiography, and planning, anthropology is always the starting point, and it is with an interrogation of anthropology's approach to time that I end. That said, these issues are bigger than anthropology and affect all scholarship of all living things.

An indication that such issues are not limited to anthropology is Jiří Wackermann's critique of metaphors of time in psychophysics (2011). Wackermann's familiarity with horological history (see 2008) allows him to point out the historical contingency of the metaphors used in the study of the psychology of time perception, and to permit him to explore alternative metaphors, such as his advocacy in thinking about psychological timing in terms of clepsydrae rather than clocks (Wackermann 2011; Wackermann and Ehm 2006). The difference between these devices is that clepsydrae indicate time by a continuous accumulation of water that reaches a threshold which marks a time unit, whereas clocks mark time through the counting of pulses—whether these pulses are mechanical or atomic. The clepsydra metaphor is much closer to how neurons function than is the clock metaphor.

So this book is best viewed as a symptom of growing discontent with our current set of assumptions about time—a discontent that is manifest not only in Wackermann's work, but in debates about the leap second policy, as well as in discussions of time in theoretical physics. My personal concern is that anthropology has lagged in these discussions. Temporal relativism is a more common theoretical principle in time metrology and physics than it is in anthropology.

I probably should mention my view of the physical sciences, since so many in anthropology, particularly cultural anthropology, are critical of any attempts to bridge such disciplinary divides. I do not accept a reductionist position, so when I refer to biology, it is not to explain culture in biological terms, but to incorporate biology into an interpretive domain. I do not see any reason to do otherwise, particularly since there is so much evidence that culture influences biology. That said, knowledge generated by the physical sciences is not sacrosanct—it should and must be criticized. I have engaged in such criticism elsewhere (Birth 2014a), and Chapter 2 of this book is not as much an endorsement of chronobiology as a lament that chronobiology is shaped by the same assumptions that shape ethnography. Even with such limitations, I have found chronobiology useful in considering the physiological consequences of global processes (Birth 2007, 2012).

Some of the chapters in this book have appeared elsewhere. Chapter 4 was originally published in 2006 as "Past Times: Temporal Structure of History and Memory" in *Ethos* 34(2), pages 192–210. The final chapter is a revised version of "The Creation of Coevalness and the Danger of Homochronism," published in 2008 in the *Journal of the Royal Anthropological Institute* (N.S.) 14(1), pages 3–20. Those chapters benefited from the comments of the reviewers, which included Gelya Frank, Joanne Rappaport, and Johannes Fabian. The final chapter also benefited from comments from Mandana Limbert, John Collins, and Alexander Bolyanatz.

There have been many conversations that have shaped this work. I've had influential exchanges about the topic of time with Axel Aubrun, Alex Bauer, Michelle Bastian, John Collins, Omri Elisha, Murphy Halliburton, Mandana Limbert, Jim Moore, Michael Northcott, Kate Pechenkina, Tim Pugh, Rob Seaman, Ken Seidelmann, Chris Sinha, Larissa Swedell, and Jiří Wackermann.

I also want to mention those venues and communities of scholars that have pushed me into new areas far beyond the study of concepts of time in Trinidad. These include the Frick Collection, the Subjective Duration colloquium, the Temporal Design workshop and the Ancestral Time project at the University of Edinburgh, the "Utopias, Futures and Temporalities" workshop in Bristol, the United States Naval Observatory, and the "Requirements for UTC and Civil Timekeeping on Earth" colloquium at the University of Virginia.

Finally, I thank my wife Margaret for her love and patience over these many years of my journey in studying time, and also for her willingness to work with me to improve this book.

PRELUDE

The Duplicity of Time

Henri Hubert, one of the founding figures in the anthropology of time, wrote, "The division of time entails the maximum of convention and the minimum of experience" (1999, 70). Barbara Adam, a leading current figure in the social scientific study of time, writes, "Time forms such an integral part of our lives that is rarely thought about" (1995, 5). The lack of thought about time can lead to unexamined assumptions, and these assumptions can mislead us when we study the ideas of time from other cultures and eras. Along these lines, A.J. Gurevich observed, "The present perception of time bears very little resemblance to that of other epochs" (1976, 230); Robert Levine notes, "Life on clock time is clearly out of line with virtually all of recorded history" (1997, 81–82); and Sacha Stern argues that assumptions about temporal regularity common today are unfounded in the study of ancient calendars and consequently lead to misunderstandings (2012). These observations suggest a problem for the study of cultural differences, namely, that a dimension of thought and experience about which we are unreflective might be shaped in ways that are culturally unusual. As I wrote in *Objects of Time*, "In thinking about the human understanding of time through the human past and across cultural differences, we have adopted a unique and artifactually mediated set of ideas as the ideal type against which all other ideas are understood and evaluated" (2012, 169). Could attempts at studying cultural differences be refracted through the unusual, even eccentric, assumptions about time derived from relatively recent developments in the European timekeeping tradition?

This is not a question of how different disciplines think about time. Andrew Abbott began his book *Time Matters* with the question "Historians cared about sequence and order. Sociologists didn't. Why?" (2001, 4). He eventually arrived at a conclusion that emphasized disciplinary differences—historians emphasize change; sociologists emphasize fixed causes (2001, 295). In effect, different disciplines have different temporalities. What are the temporalities of anthropology, however? If one accepts Malinowski's dictum that anthropology studies the "native's point of view" (1961, 25), then one might conclude that anthropology merely relates the temporalities of the people it studies, but the crisis of representation in the 1980s (see Clifford 1983; Clifford and Marcus 1986; Marcus and Cushman 1982) revealed that claim to be dubious. Whereas Abbott is interested in what the dominant temporality should be for theory and method, the question raised here is a bit different: What temporalities get in the way of describing and understanding the diversity of human thought and behavior?

Anthropology is among those disciplines one might think are equipped to address cultural diversity and time, and to become aware of its own temporal assumptions. Yet, anthropology has struggled with this issue. In *Time and the Other*, Johannes Fabian raised the issue of "anthropology's problem with Time" (2002, 60). Unfortunately, while he pointed out the problem, anthropology has not really addressed it. Rather than wrestling with the time problem, much of anthropology instead offers a greater "historical perspective" than it did previous to Fabian's book. While this addresses the temporal warping caused by the rhetorical construction of the ethnographic present, it does not address the underlying cultural logics that shape Western scholarship's conceptualization of history. Indeed, history is as subject to these temporal logics as anthropology, so a substitution of history for the ethnographic present is simply a substitution of one manifestation of these logics for another. Indeed, the problem with time is not just within anthropology, but is a problem of all European-derived post-Enlightenment scholarship. Fabian was too modest in his project.

Carol Greenhouse's book *A Moment's Notice* is bolder than Fabian's in its trenchant criticism of anthropological temporal biases, because rather than limiting herself to the temporal framing of ethnographic representation, she tackles the biases that seep into all ethnographic practice. She notes numerous ways in which anthropology's naive approach to the topic of time distorts ethnographic representation: she points out the tendency of anthropologists to assume that without time-keeping technology, ideas of time emerge from cultural interpretations of nature (1996, 40); she describes how anthropologists assume that a lack of cultural emphasis on duration indicates a lack of concern about time (1996, 41); and, finally, that "the indifference to time mentioned by some ethnographers is generally taken to indicate an absence of temporal constructions altogether" (1996, 46). According to Greenhouse, all of these are examples of how anthropology has deviated from Durkheim's emphasis on time as social in favor of an assumption about a "real" time against which different cultural notions can be evaluated (1996, 46–47). She then notes that anthropological discussions of social time "proceed from analogies to the mechanical clock, as if the clock were itself a materialization of some universal time sense" (1996, 47). The clock is a European cultural creation. As I have argued elsewhere, it is the clock and Gregorian calendar that are exotic and unusual ways of representing time—"the exceptions of human history that have become the rule" (2012, 170). Postill has chastised anthropologists for not studying the spread of calendar and clock time throughout the world (2002), but to his critique can be added the semiconscious absorption of clock and calendar time by anthropologists when describing temporalities. Anthropology has even been an agent in spreading clock and calendar time by means of its representational strategies.

Recently, John Collins (2015) has argued that there is a tendency among social scientists to fix the ontological status of cultural representations without giving due attention to how those representations are fluidly negotiated, created, and deployed. In the process of doing so, social scientists often insert their own cultural biases that then influence how the cultural representations are described. Based on Greenhouse's observations, this is particularly true with the study of cultural concepts of time where European-derived temporal concepts have filtered ethnographic representations in ways that are subtly distorting—whether it be in the form of time-allocation studies, or representing a daily round of activity as parallel to a clock, or treating how people chart annual cycles as if they used European calendrical logics. In effect, the biases that Gurevich, Levine, and Stern observe can distort ethnographic description and historical scholarship. Moreover, there is growing evidence that the spatial metaphors that European-derived scholarship often uses to discuss time are not pancultural— namely, that there are languages in which spatial metaphors are not used in this way (Sinha 2014a, b; Sinha et al. 2011).

It Can Be Noon at Two O'Clock

Even European timekeeping used to be different from our current naturalized assumptions about time. To demonstrate this, I shall discuss Derrida's analysis of the story "Counterfeit Money" by Charles Baudelaire. Derrida is regarded as a major figure in unsettling the connection between representations and their meaning. His concept of deconstruction has been applied widely to reveal hidden assumptions and contradictions in texts ranging from the literary to the scientific. Yet, Derrida was not immune to the influence of his cultural milieu. His discussion of "Counterfeit Money" reveals his assumptions about time, and the gulf between him and Baudelaire. In the book *Given Time: I* (1992), Derrida addresses the relationship of time to giving in this story of Baudelaire's. To focus on the problem of giving, Derrida cannot allow time to be a self-referential signifier with multiple, relevant meanings. So when discussing Baudelaire's use of a French idiom—"But into my miserable brain, always concerned with looking for noon at two o'clock (what an exhausting faculty is nature's gift to me!)" (Baudelaire quoted in Derrida 1992, 32)—Derrida writes:

> At no *given moment*, and no *desired moment* [moment voulu] can one reasonably hope to find, outside any relativity, noon at two o'clock. This contradiction is the logical and chronological form of the *impossible* simultaneity of two times, of two events separated in time and which therefore cannot be given *at the same time*. (1992, 34, emphasis in original)

It *seems* quite obvious that it cannot be noon at two o'clock, but it is only obvious *within* a twentieth-century post-Enlightenment understanding of time as a uniform commodified duration. Derrida does not recognize clock time for what it is: a self-referential representation (see Birth 2012). It cannot be noon at two o'clock only because that is what we were taught in grade school when we learned much of what we know about time. Then again, as Elias observes, "One of the difficulties in investigating time is that people are as yet little aware of the nature and functioning of the symbols they have themselves developed and constantly use. They are therefore always in danger of losing themselves in the undergrowth of their own symbols" (1992, 29).

I know from my French friends that the meaning of the colloquial phrase "looking for noon at two o'clock" is to create a problem where there is none, but I shall purposefully complicate that colloquial meaning in order to demonstrate something Derrida ignores in his discussion: that

Baudelaire wrote his story in a context with different attitudes and ideas about time than when Derrida generated his analysis. In a sense, by problematizing "looking for noon at two o'clock," I am "looking for noon at two o'clock" by looking for noon at two o'clock.

In looking for trouble, I ask: What if time is duplicitous? What if underneath the seemingly obvious fact that it cannot be noon at two o'clock is a set of multiple meanings that permit noon at two? In fact, duplicity in the measure of time is at the very core of modernity's timescales. Some might find the explanation of this tedious, but understanding the technical side of how time is produced sheds light on how self-referential, and potentially duplicitous, the representation of time can be.

The International System (SI) second, the global standard unit for measuring time, is defined by periods of cesium atoms. That said, while the second is defined in terms of periods of cesium, the best clocks today are not cesium clocks, so the cycles of cesium equivalent to a second are actually measured by the periods of rubidium or hydrogen. The standard definition of the second is different from most people's understanding of what a second is: The SI second is *not* defined as a fraction of a mean solar day, because Earth's rotation is too irregular to serve as a measurement standard. When the SI second was originally defined, it was in reference to a fraction of a tropical year (CNRS 1950, 129), because the duration of Earth's orbit around the Sun is far more stable than Earth's wobbly rotation.

This SI second is not a fraction of the current year as measured by atomic clocks, however. This fraction of a year that defines the second (as represented by cycles of atoms) is roughly equivalent to the average length of 1/86,400 of a solar day as calculated by Simon Newcomb in 1895 for January 1 of the year 1900. Newcomb used astronomical records from the eighteenth and nineteenth centuries to determine the average rotation of Earth and the duration of that rotation. He then projected his calculation forward to 1900. Modern metrologists think that Newcomb's data set is such that his calculation more accurately refers to a day sometime in 1820, however (Nelson et al. 2001, 509).

In sum, the unit of the second is not a simple matter to define. On the contrary, it is a polysemic sign that a global bureaucracy concerned with measurement attempts to constrain to a single meaning. Such attempts at constraint are undone by their own technological efforts to improve the definition of what a second is. The attempts by scientists to improve the definition result in the proliferation of new and old definitions used by the general public and embedded in previous clocks, watches, and electronics. To date, these differing definitions have not been great enough for most people to notice, but as more and more technologies rely on greater and greater precision, legacy definitions will become a greater problem.

So if a second is precisely defined, then a minute, an hour, and a day are precisely defined, right? This is true of the minute and the hour, but not of the concept of the day. 86,400 SI seconds never vary; rotational days do. Since Earth is a rotating object moving through space, one rotation in relationship to the Sun usually does not equal 360 degrees. This rotational period varies further because Earth is tilted, and the orientation of the tilt to the Sun affects the duration of a day. These variations led to the calculation of the mean solar day—an average length of day throughout the year. Moreover, since Earth's surface is mostly water, and the water gets pulled by the Moon to form the tides, there is constantly a bulge moving around Earth as it rotates. This bulge of water generates a braking influence on its rotation that results in that rotation gradually slowing over time. Because of this effect, Nelson et al. point out that "the present length of the mean solar day is about 2.5 ms longer than a day of precisely 86,400 SI seconds" (2001, 75).

So is a day an 1820 mean solar day, 86,400 SI seconds, a mean solar day in the present, or an apparent solar day? Actually, the day has not been officially defined as a unit of measure (see Seaman 2014).

The duplicity of time does not end with the relationship between the second, the day, and Earth's rotation. The definition of the SI second is used to constitute Coordinated Universal Time (UTC). UTC is a weighted average of times indicated by over 300 atomic clocks. There is no master clock. Each month, the International Bureau of Weights and Measures (BIPM) calculates UTC and reports the value of UTC for the previous month in the publication *Circular T*. This periodical not only contains the previous month's values for UTC but it also records the deviations of the times of all the laboratories that contribute time data for UTC's calculation. This means that the UTC which the time services of these laboratories distribute is not authentic UTC, but an estimation of UTC (Arias et al. 2011, S148–150; Levine 2001, 54). It is this laboratory-specific estimation of UTC that goes out to most of our electronic devices. Thus, even UTC is duplicitous, although for most practical purposes, the difference between the BIPM calculation and the time that is distributed is too small to matter.

So, was Baudelaire's sense of noon 43,200 seconds after midnight (probably not, because his clocks and watches were likely not that accurate), or when his clock indicated 12:00 p.m., or when a public clock struck twelve, or the general period around midday, or when a shadow was at its shortest during the day, or when a shadow pointed due north or struck a noon mark on some windowsill? There are multiple noons and multiple twos. Derrida's comment should be changed from "outside any relativity" to "outside any consciousness of the semiotics of time, it cannot be noon at two."

Built on this duplicity of time is a duplicity more familiar in social science—the duplicity of time in defining capitalist labor relations. Sometimes the artistry of social relations requires managing different times. In a context such as rural Trinidad where industrial and agricultural labor relations meet and collide, one finds a consciousness of multiple times that Derrida lacks. In Trinidad, when I inquired about the expression "Any time is Trinidad time," one man said, "You'll tell the person who comin' to meet you to go come at one o'clock and their one o'clock might be half past two in the evening, and when they come they will still tell you they on time." On the surface this seems quite commonplace and different from the sensibility in Baudelaire's phrase, but the logic behind the saying "Any time is Trinidad time" that allows one to treat eight o'clock as if it were seven extends to other contexts, and most interestingly in relationships between workers and their employers. To frame this point, it is worth quoting an elderly Trinidadian man who recalled how time was indicated in plantation work: "In the instances they blow a conch shell to tell you, well, in ten minutes we all must bear up to enter the field and at dismissal time, at four o'clock, they blow the conch shell again, and you know it is dismissal time." This centralized control of time is a facet of labor relations. E.P. Thompson offers testimony from witnesses to conditions at Braid's Mill: "'...[I]n reality there were no regular hours: masters and managers did with us as they liked. The clocks at the factories were often put forward in the morning and back at night, and instead of being instruments for the measurement of time, they were used as cloaks for cheatery and oppression'" (quoted in Thompson 1967, 86). The implication of this is that the employer could make four thirty or even five o'clock into the dismissal time of four o'clock.

When workers carry their own watches, it is not merely a matter of keeping the employer honest, but it allows the possibility of the workers themselves creating duplicitous time. Where I did my fieldwork in

Trinidad, the conditions were ideal for this. In both plantation and roadwork, workers were distributed over large areas and often supervised by a foreman who was promoted from the ranks of the regular workers, but whose sentiments were still more with the workers than with the employer. This foreman was in charge of the dismissal time from work and obviously could become popular with the workers if he allowed the workers to go home early. If management suspected that a foreman was doing this, they would send a supervisor out to the work crew around dismissal time to check to see if they were working. In this game of cat and mouse, the foreman had to be sure to dismiss workers at the correct dismissal time according to his watch. This led to exchanges between supervisors and workers about the correct time. As I discussed in *Any Time Is Trinidad Time* (Birth 1999), foremen developed a trick of being able to change the time on their pocket watches while the watches were still in their pockets. This would allow a foreman to set his watch to that of the supervisor at the beginning of the day, end the work day 20–30 minutes early, and adjust the watch to indicate the correct dismissal time when a supervisor showed up at the work site after the work crews had gone home (1999, 103–104).

Returning to Derrida, rather than viewing noon at two o'clock as an "obvious" contradiction, thereby a priori prohibiting any inquiry into other possibilities, one should be open to the ways in which it can be noon at two o'clock. To do this requires attention to the flexibility of noon and the artificiality of the clock. After all, for all the emphasis that has been made on the arbitrariness of signs, little of this discourse has been applied to the arbitrariness of clock time as a sign.

Derrida's sense that noon at two o'clock is an obvious impossibility is the product of a temporal consciousness that seems to have emerged relatively recently as a result of the acceptance of systems of uniform timekeeping. Elias points out that the current emphasis on uniformity and continuity "runs counter" to both physics and the thinking of our predecessors (1992, 39–40). In fact, Baudelaire's life span overlapped with the period when the foundation for temporal uniformity was being laid but temporal uniformity was not yet adopted. This was the period after what Macey (1980) dubs the *horological revolution*—a period in the eighteenth century when the precision of timepieces coupled with the precision of astronomical observation ushered in a new era of accuracy in timekeeping. As a result, mechanical clocks and mean time came to be preferred over direct observation of the Sun. One reason for the preference was that clocks were easier for travelers to use than sundials. To use a sundial properly, one must know one's latitude as well as be able to orient the dial to a north–south axis. A watch required less knowledge to use as one traveled. One could simply keep it set to local time by adjusting it to local bells that chimed the hours or to noon marks or public sundials that could be found in many towns and cities. There were no time zones at this period, so even a short journey east or west could result in having to adjust one's watch.

Baudelaire lived at the end of this shift from sundials to watches. It is likely that he was even aware of French Revolutionary time, which replaced traditional time with decimal hours (Shaw 2011; Zerubavel 1977). In this time, noon was five o'clock. This is readily seen in period watches that were made to allow one to convert decimal time to traditional time.

Baudelaire's life also unfolded somewhat parallel to that of Edgar Allan Poe, a writer who often wrote about the theme of time and whose work was translated by Baudelaire into French. Baudelaire, in effect, worked during an era in which time reckoning was shifting from the individualistic timepiece to the synchronized watch, and he translated Poe's texts in which the theme of time reckoning looms large and usually in connection to themes of foreboding (as in "The Raven"), terror ("The Pit and the Pendulum" and "The Tell-Tale Heart"), suspense ("The Masque of the Red Death"), the emotional range from rapture to melancholy to despair ("Bells"), or the foolishness of those who do not understand the arbitrariness of clocks and calendars ("The Devil in the Belfry").

This last story is particularly interesting in terms of period consciousness regarding the duplicity of time. "The Devil in the Belfry" portrayed how the people of Vondervotteimittiss (Wonder-what-time-it-is) lived by their impressive town clock. This clock was managed by the belfry-man, who is the most important man in the village, or, as Poe writes, "He is the most perfectly respected of any man in the world. He is the chief dignitary of the borough, and the very pigs look up to him with a sentiment of reverence" (1975, 739). One day, just before noon, a "diminutive, foreign-looking young man" appeared carrying a *chapeau-de-bras* and a large fiddle. He danced into town to the annoyance of the residents of Vondervotteimittiss—"But what mainly occasioned a righteous indignation was, that the scoundrelly popinjay, while he cut a fandango here, and a whirligig there, did not seem to have the remotest idea in the world of such a thing as keeping time in his steps" (1975, 740). This man "pigeon-winged himself right up into the belfry" (1975, 740), and as the great clock began to strike noon, he began to bludgeon the belfry-man with the fiddle and the *chapeau-de-bras*. Meanwhile, the townspeople diligently counted the chiming of the clock's bells. After the clock struck noon, the story concludes:

> "Und dvelf it is!" said all the little old gentlemen, putting up their watches. But the big bell had not done with them yet.
> "Thirteen!" said he.
> "Der Teufel!" gasped the little old gentlemen, turning pale, dropping their pipes, and putting down all their right legs from over their left knees.
> "Der Teufel!" groaned they, "Dirteen! Dirteen!!- Mein Gott, it is Dirteen o'clock!!"
> Why attempt to describe the terrible scene which ensued? All Vondervotteimittiss flew at once into a lamentable state of uproar.
> "Vot is cum'd to mein pelly?" roared all the boys—"I've been ongry for dis hour!" (1975, 741)

In this story, among Poe's targets for his wit and mockery are those who wonder what time it is and who are obsessed with the clock. He portrays these folk as so focused on the clock that they cannot recognize noon. They make the belfry-man the most important man in the world. But a scoundrel who cannot keep time when he dances demonstrates that thirteen o'clock is possible, and that it is possible for it to occur at noon.

It is not surprising, then, that time is a theme that recurs in Baudelaire's writings as does his ambivalence about clock time. Baudelaire's life unfolds between the emergence of reasonably accurate watches and the enshrinement of national times and time zones in law. It is a period when each town has its own local time, and even public clocks within the same city can strike the hour differently. The combination and contrast of different times and timescales is a powerful generator of meaning rather than simple contradiction or irony.

The difference between passages about clock time from before Baudelaire's birth and after his death reveals the movement from multiple to uniform times. Baudelaire is a writer during the period of the muddled middle of timekeeping—between diversity of times and clock performance and uniformity and grades for clock performance. For instance, in his 1711 *Essay on Criticism*, Alexander Pope wrote:

> 'Tis with our *Judgments* as our *Watches*, none
> Go just *alike*, yet each believes his own. (1970, 4, emphasis in original)

As an aside, in Samuel Johnson's collection *The Works of English Poets*, he lists the date of Pope's poem as 1709 with an asterisk that leads to the note: "Mr. Pope told me himself that the 'Essay on Criticism' was indeed written in 1707 though said 1709 by mistake. J. Richardson" (1779, 87).

This particular phrase of Pope's has remained popular in English literature and has been deployed in a variety of ways, ranging from a guide for military surgeons (Hamilton 1787, Vol. 2, 179) to an article on circadian rhythms of birds and plants in the *New York Times* in 1998 (Freeman 1998, 4). While the phrase suggests the great extent to which opinions differ, its survival in an age of a coordinated global time system that is managed to a level of accuracy of one second every 60 million years seems ironic. If opinions varied as much as the clocks and watches of the early twenty-first century, then opinions would vary hardly at all! So while Pope's turn of phrase poetically indicates varied opinions, it does so solely on the consciousness that different watches can indicate different times. If one accepts Pope's representation of the extent to which opinions vary as an indication of the extent to which early eighteenth-century watches varied, then watches used to vary quite considerably in the times they indicated. Milham reports that early clocks and watches were only accurate to around two hours per day (1947, 226). Before Baudelaire's lifetime, for it to be noon at two *of the clock* one had to use a clock—one could not achieve such confusion with a sundial.

In fact, in the English-speaking world of the eighteenth century and later, statements of another time at noon were commonplace—a search using the digital archive *Eighteenth-century Collections Online* produces many results along these lines: In his memoirs, General Thomas Fairfax wrote of one of his battles, "And here did the first fifth continue from eleven o'clock at noon, till five at night" (1776, 29); and Mrs. Catherine Jemmat wrote in her memoirs, "Yesterday about two o'clock at noon, I came to drink a pint of beer at a public-house" (1765, 139). Such minor examples are numerous, but one also finds the following two cases. The first comes from a letter to *The Monthly Magazine and British Register* from 1798:

> Sir,
>   I have just been reading in your Magazine for July, an excellent paper on "*Progressive lateness of Hours kept in England,*" and heartily concur with the sentiments offered there on this growing folly …
>   I believe that four o'clock is the latest dinner-hour in the memory of the oldest fashionables now living. This was soon altered to five, which, with some, is still nominally the hour: I say nominally, for cards of invitation, like the beauties of Eastern writing, are not to be interpreted literally; and *five generally means,*

> *and is fully understood to mean, any time between six and seven* ... One night at the opera, when the last dance was finished, I heard Mr. B— ask Lord D— to go home and take pot-luck with him, which the latter declined, owing to a previous engagement to dine with a select party, as soon as the Duke of Bedford's motion was got rid of! This Mr. B—, however, is a sort of wag—a plain country gentleman, who eats his mutton chop quietly at *two o'clock at noon*, and afterwards sups at his lady's midnight dinners. (1798, 97, emphasis added)

The second case comes from *The Spectator*, a daily periodical written by Joseph Addison and others in the early eighteenth century:

> The hours of the day and night are taken up in the cities of London and Westminster, by people as different from each other as those who are born in different centuries. Men of six o'clock give way to those of nine, they of nine to the generation of twelve, and they of twelve disappear, and make room for the fashionable world, who have made *two o'clock the noon of the day*. (1744, 226, emphasis added)

Noon is thus sometimes a period around midday, and sometimes the zenith of the Sun, and sometimes twelve o'clock. The "fashionable" seem to be fashionably late with time, commonly making five o'clock into six or seven in the evening, and two o'clock into noon. There is a consciousness of the arbitrariness of time.

It is a small conceptual difference between making two o'clock noon and looking for noon at two o'clock as Baudelaire's protagonist does. The fashionable eighteenth-century English could find noon at two, but what of Baudelaire's Parisian? For him, this ability was a curse, and led to moral musings on his friend's behavior. Contrary to Derrida, the tension is not between noon and two, but between the curse of looking for the fashionable noon at two and a moral sensitivity toward beggars that is not fashionable.

This discomfiting nature of time was later suppressed. In contrast to Pope, in *Around the World in 80 Days* (a tale based on Poe's story "Three Sundays in a Week"), Jules Verne describes his protagonist, Phileas Fogg, as following a daily schedule according to "chronometrically set times" (1995, 9), and Fogg's abbreviated interview in which he hires Passepartout as his valet is instructive:

> "What time do you make it?"
> "Eleven twenty-two," replied Passepartout, pulling an enormous silver watch from the depths of his waistcoat pocket.

"Your watch is slow."
"Pardon me, sir, but that's impossible."
"You're four minutes slow. It is of no consequence. What matters is to note the difference." (1995, 11)

The reader learns from Passepartout that his watch "never loses more than five minutes a year. It's a genuine Chronometer!" (1995, 35). Yet, instead of setting his chronometer to a known meridian, Passepartout must align his reckoning of time with that of his master.

Whether or not the idea of a "genuine chronometer" is meaningful depends on the audience. Connoisseurs of fine watches might appreciate the statement and the irony that it is the servant who carries the chronometer. Since 1973, makers of fine spring-oscillator watches have sought chronometer certification from the Contrôl Officiel Suisse des Chronomètres (COSC). Such certification demonstrates the accuracy and reliability of the watch. Those, like myself, who are content with (and can only afford) cheap watches tend not to aspire to purchase a watch that is a genuine chronometer. Yet, this certification indicates that two watches that show the same time, an expensive spring-oscillator certified chronometer and a cheap electronic watch, suggest a profound difference in status between their owners.

The contrast between Pope and Verne suggests that innovations in timekeeping that emerged during the eighteenth and nineteenth centuries have made all watches "go alike."

Since clock time was a signifier in flux, as the global time system evolved and before there were international agreements seeking temporal uniformity, there were many different ways it could find noon at two.

THE DUPLICITY OF NOON

In current colloquial English usage, noon is twelve o'clock during the daylight hours. In astronomy, apparent noon is when the Sun is at its zenith at a particular meridian. The term *noon* has its origins in the canonical hours of the Christian Church.

Originally known as Nones, noon is the ninth hour after sunrise, but in more practical terms, Nones was when the Sun was halfway between its zenith and its setting. In some parts of Europe, Nones began to drift earlier in the day, until reaching midday, thereby creating the conflation between the canonical hour and the time of the Sun's zenith in the term *noon* (Dohrn van-Rossum 1996, 31). Since the canonical hour of noon/ Nones was in the midafternoon, this gives some irony to the translation of Derrida's French into the English "impossibility of noon at two o'clock." In French, this confusion of noon and Nones does not exist: *midi* is midday, and Nones is the canonical hour.

Until as late as 100 years ago, throughout much of Europe, public bells chimed the canonical hours rather than mechanically determined hours (Corbin 1998, 111). The canonical hours were developed around seven periods of reciting components of the Daily Office. The daily cycle of readings, canticles, and psalms that the religious recited may have begun as early as the first century (Taft 1993, 13). The legitimacy of the seven periods came from Psalm 119:164, as well as from the story of the Passion of Jesus Christ, one of the few stories in the Bible with a detailed representation of diurnal time. These passages from the Bible led to the monastic hours of Matins, Prime, Terce, Sext, Nones, Vespers, and Compline. Each monastic hour was linked to a place in the cycle of daylight, and consequently varied throughout the year as the balance of daylight and nighttime shifted.

Since church bells were among the earliest and most public of timekeepers, and since these bells chimed monastic hours and not mean hours, once clocks and watches that worked on the basis of mechanical mean time were developed, there was a clear discrepancy between church bells and mechanical timepieces. Slowly, throughout Europe, the chimes were synchronized with mean time, with some of the last conversions occurring in the early twentieth century.

The phrase "noon at two o'clock" is a pivotal point in Baudelaire's story—it is the moment when the narrator begins to reflect on the dubious morality of giving a beggar a counterfeit coin. Derrida sees an impossibility, but the dual meanings of *midi* in French as midday and lunchtime compounded with the multiple associations of noon in English do not create a contradiction, but a dense web of association between being fashionable, eating, fasting, penance, and religiosity. As long as Derrida takes the time as obvious, it is not possible to recognize the alternate possibilities, much less to challenge modernity's assumptions about time.

Post-Baudelairian Noons at Two

This discussion has raised the ways in which it might be noon at two o'clock in order to reveal the irony of Derrida accepting an artificial, constructed, and arbitrary time so as to allow him to create the obviousness of a contradiction. This is ironic, because rather than engaging in deconstruction, Derrida is affirming a false consciousness of a cultural construct of modernity. This discussion also serves as a means of pointing out the contrivance of modernity's temporal uniformity. The point is to reveal the extent to which we accept as obvious a sense of time that is historically and culturally contingent. The idea of homogenous empty time criticized by Benjamin (1968), yet widely accepted as a condition of modernity (Anderson 2006; Asad 2003; Taylor 2007), does not have the historical depth and naturalness that is often attributed to it, and as later chapters will show, it should not have a complete grip on consciousness.

This cultural emphasis on temporal uniformity is antagonistic to other temporalities. In his *Principia*, Newton argued in favor of the use of absolute temporal uniformity over observed motions. For him, true time was this absolute and uniform time, not the colloquial unit of hour, day, month, or year. In effect, Newton argues that science needs to be based on a true time that is distinct from the rhythms and temporalities people experience. In an odd turn of phrase, he writes, "Possibile est, ut nullus sit motus aequabilis quo Tempus accurate mensuretur" [It is possible that there may be no uniform movement by means of which to measure time accurately] (1714, 7). The implication is that it might not be possible to achieve truly uniform units of time because all the processes and cycles in the world are too irregular. Somehow, this unachievable (to Newton) idea of time is the dominant idea of time by means of which everything is studied.

The most significant blindness of Derrida in his *Given Time: I* is an experience of something even more pronounced than noon at two that he must have had with great frequency. After all, in his later years he taught at the University of California at Irvine. If he had ever had a telephone call between California and France, he would have had to recognize a time difference of nine hours, in effect, noon in California and nine in France. He would have lived nine at noon! The impossibility he proclaims for noon at two, then, is an undue deference to clock time that ignores all the ways in which it can be two times at once.

This time difference is not directly related to Baudelaire's noon and two, but it does undermine the impossibility and contradiction that Derrida attributes to Baudelaire. Instead of reading Baudelaire as making a comment on impossibility, it is possible, contra Derrida, to read Baudelaire as emphasizing equivocality and duplicity—the duplicity of time and the gift of counterfeit money.

WHAT IS TIME?

Marramao points out that "[t]ime places itself at the crux of the relation between daily experience and its representation" (2007, 39). The problem with Derrida's reading of Baudelaire's comments about time is that Derrida takes for granted the merger of experience and representation. For Derrida, the definition of time by the motion of the clock is unquestioned and even naturalized. This seems to be part of the modern condition of social science. In physics, on the other hand, the clock does not go unquestioned. Einstein makes the relativity of time and motion a central concern, while Derrida emphasizes that it cannot be noon at two o'clock "outside any relativity." In other words, outside of what is known about the physical nature of time, it cannot be noon and two o'clock at the same time. In Derrida's imagination of time, unquestioned and naturalized cultural signifiers trump physics. Yet, if they are cultural signifiers, they could be otherwise. Outside of assumptions of European-derived forms of modernity, it can be noon at two.

Why devote this much attention to Derrida's *Given Time: I*? The point is to unsettle naturalized ideas of time so that it is easier to recognize their misapplication. The chapters that follow explore such misapplications in ethnographic representation, chronobiology, discussions of labor-time, chronology, and the imagination of the future. The challenge is to overcome such cultural bias to explore temporalities in all their forms.

CHAPTER 1

(Hegemonic) Calibrations in Anthropology

About her first fieldwork, Hortense Powdermaker recalled, "The second day in Lesu, I went swimming with my watch on my wrist!" (1966, 77). This is an odd observation to record for posterity. In an era when anthropologists sought the "primitive" with the expectation of finding prehorological thinking, the anthropologist Hortense Powdermaker not only was explicit about her timepiece, but she also mentioned how briefly it survived in the field. Maybe she was emulating her mentor, Bronislaw Malinowski, who also reported ruining his watch after a good dunking early in his fieldwork (1989, 27). Maybe baptizing a watch was a part of a secret ritual of fieldwork at the time.

The anthropology that emerged in the 1920s, the anthropology in which Powdermaker placed herself at the time, emphasized function. Society was seen as a means of meeting the biological and psychological needs of its members. The main proponent of this approach, Bronislaw Malinowski, argued in favor of long-term field research to ascertain local needs and how culture met these needs, but he developed a style of ethnographic representation "resolutely in the present tense" (Sanjek 1991, 613). This "ethnographic present" created a sense of timelessness in anthropological discussions of its subjects (Fabian 2002). Ruining a watch seems to be a suitable symbol for the creation of the timeless ethnographic Other.

Yet, Powdermaker and Malinowski were careful to keep track of time while conducting studies even though their final analyses would not emphasize the passage of time. Powdermaker stated that she used a calendar; Malinowski's diaries were not only ordered according to the

Gregorian calendar but they contained frequent references to clock time. (Did he replace his watch or convert daily cycles into estimates of clock time?) Keeping track of time was not merely for the sanity of the ethnographer but it was also an important component of ethnographic field technique. *Notes and Queries*, the definitive guide to fieldwork in this period, instructed its users to learn the local measures of time (Freire-Marreco and Myres 1912, 235–236). This guidance seems based on an earlier set of questions crafted by Frazer and published in the *Journal of the Royal Anthropological Society of Great Britain and Ireland* (1889, 435).

Malinowski struggled with this goal, particularly with regard to the Trobriand "calendar." Among the Trobrianders he interviewed, there was no consensus about the number of moons in a year (1927, 209). In his diary, he writes "long after parleys" with two informants: "At last the mystery of the months is being clarified" (1989, 30). In a later article, he observes that Trobriand calendars did not work like the Gregorian calendar. Still, the annual cycle he included early in his classic *Argonauts of the Western Pacific* was organized according to the Gregorian calendar (1961, 16). Early in the history of ethnographic representation, the Gregorian calendar became an unacknowledged lens through which to view cultural practices of time reckoning and an obstacle to achieving what Malinowski called "the natives' point of view" (1961, 25).

The use of the Gregorian calendar over other calendars privileges it over other ways of thinking about time. It also suggests that time was naturalized in the anthropological mindset in ways that were not true of other topics for research, such as kinship. In British social anthropology, kinship became the central focus. This is because W.H.R. Rivers argued that kinship not only reflected cultural differences but that it indicated different social structures, with kinship relationships being the "bedrock" of any social structure (1911, 394). As a result, according to Rivers, the anthropologist had to be sensitive to the subtle differences between kinship systems: "In nearly all people of the lower culture these [kinship relationships] differ so widely from our own that there is the greatest danger of falling into error if one merely attempts to obtain the equivalents of our own terms" (1910, 3). Anthropology worked to represent kinship in terms of cultural diversity rather than through the lens of European kin terms. There was (and is) no equivalent to the genealogical method developed by Rivers for studying concepts and practices related to time. As a result, these concepts and practices tend to be refracted through European-derived categories associated with calendars and clocks. In contrast, when non-Gregorian calendars were ethnographically represented, they were often converted into Gregorian terms; when times of day were represented, they were converted into approximate clock times (see Evans-Pritchard 1939; 1940, 94–104).

To Malinowski's credit, he struggled against this tendency, even though he ultimately succumbed to it. Since lunar time reckoning was important in the Trobriand Islands, Malinowski initially assumed that they would name 13 lunar months. The reason for 13 months is that the 12 lunar months are shorter than a solar year, and yet the Trobriand Islanders' sense of a year was closely tied to the spawning of sea worms. As a result, he seemed to assume that the Trobrianders would have to occasionally add an extra month to their count—a 13th month. This is an example of how approaches that assume all ways of reckoning annual cycles work like calendars lead to misunderstanding (see Stern 2012). As a result of this problem, Malinowski quickly faced trouble: "[A]lthough there existed names for various moons, there were not thirteen, and it was difficult to find out how many names there were. From some I obtained ten, eleven, or twelve, and sometimes, under pressure, thirteen, but it was clear that there was no universally known figure" (1927, 209). In reflecting on this difficulty, he offered the following insight:

> I took for granted that when natives name moons and count them, this is the purpose of time-reckoning ... and I assumed that the whole scheme was a system of time co-ordinates. The correct procedure, however, would have been not to assume a given use or function in the scheme, but only to enquire into it ... *In other words, the next step should have been to divest myself of our mental and cultural habits.* We name moons for calendar purposes, and we use the calendar to divide and count time and to fix dates; and with us the whole system with its many ramifications is a system of time co-ordinates. This, however, does not mean that a similar system obtains in simpler cultures. (1927, 209, emphasis added)

Malinowski then revisited his material and concluded that the Trobriand Islanders only had use for 10 or 12 month names, and therefore did not have, or need, any more.

Malinowski's trouble and his astute conclusion that not all calendar systems work like European-derived calendars was not left unchallenged. Anthropology did not give up its Gregorian proclivities without resistance. In 1939, Leo Austen, the resident magistrate of the Trobriand Islands, published a new analysis of the Trobriand "calendar" in which he argued that there were, in fact, 13 named months. He admitted, "I began with the idea that there were thirteen native moons in each year." He then argued that the moon names roughly corresponded to garden periods, but that this still led to confusion, particularly when there was a 13th moon in the year. As a result, he advocated that the garden times should be "taught by European months and dates" (1939, 251).

Edmund Leach (1950) then reanalyzed the differences between Austen and Malinowski. He pointed out that Austen and Malinowski agreed on ten of the month names. The crux of Leach's analysis is that one of the month names, *Milamala*, does not refer to a single month, but to a period of four months (1950, 253). This allowed Leach to argue that there really were only 10 month names, but that with 10 names, 13 could be covered when necessary.

In contrast to Rivers' genealogical method, which emphasized using the bare minimum of kin terms to represent cultural differences, the debate about the Trobriand calendar was driven by an anthropological bias that believed that the number of named months would correspond to the duration of a year—that there must be a calendar system that would work like systems familiar to Europeans, whether it be the Jewish system with intercalary months or the Gregorian/Julian system that emphasized the solar year. Malinowski's eventual conclusion that this was not necessary is likely the best interpretation even though it did not satisfy Leach. When Turton and Ruggles attempted to document the calendar of the Mursi in Ethiopia, they discovered that while there was agreement on the first month of the year, there was general disagreement later in the year—hence the title of their article "Agreeing to Disagree" (Turton and Ruggles 1978). The ethnographic research on the Trobriand Islanders and on the Mursi suggests that European-derived calendrical logics are not the only way to imagine time, but one should also resist the conclusion that cultural logics that are little interested in reconciling units of months and years are less advanced than those logics that address this concern. To make such an assumption would require labeling modern astronomers as primitive since they eschew the concept of the month in favor of day counts and Julian years.

In 2011, there was another example of the creation of a radical alterity on the basis of concepts of time. There were media reports of a lowland Amazonian group, the Amondawa, who, according to the *Daily Telegraph*, had "no calendar and no concept of time" (Alleyne 2011). This conclusion was a distorted interpretation of research by Sinha et al. (2011), which challenged the claim that ideas of time are metaphorically developed from the experience of movement in space—a phenomenon known as linguistic space-time mapping. In English, examples of this would include "The future is ahead of you," or "The weekend is coming." Sinha et al. argued not only that the Amondawa exhibited no space-time mapping, but also that they had no calendar system. This was not an argument that the Amondawa had no concept of time, however. On the contrary, it was an argument that Amondawa time concepts were different from European ideas and not easily represented through the lens of European time reckoning (Sinha et al. 2011; Sinha 2014a, b)—the point I am trying to emphasize here.

Sinha et al.'s data on Amondawa ways of discussing time revealed ways of indicating the past, present, and future; the seasons; the parts of a day; and stages in the life cycle. The lexicon for discussing time does not make any spatial references. In their discussion, Sinha et al. state, "We would strongly disavow any interpretation of the data that we present that would exoticize the Amondawa by suggesting that they are a 'People without Time'" (2011, 160). Since Amondawa language does not use the same metaphorical field that many languages use to discuss time, and since it does not refer to the artifacts of clocks and calendars, it is difficult to translate Amondawa time ideas into English without adding such metaphorical material. It is unlikely that the Amondawa are the only group that would pose such a problem. Still, Sinha et al.'s attempt to challenge Western philosophical assumptions about the nature of time were twisted by media reports away from this intent and toward a simpleminded documentation of otherness—of timeless "primitives." Moreover, this work has had little influence on the anthropology of time, despite its importance.

Carol Greenhouse observes, "For the most part, anthropological studies of social time have proceeded from the double assumption that linear time—'our' time—really *is* our time, and really is *real*" (1996, 2, emphasis in original). As she points out, this double assumption leads to error in ethnographic representation, and error in privileging one cultural set of ideas above all others. Anthropologists have debated whether societies view time as linear or circular (Bloch 1977, 1979; Bourdillon 1978; Howe 1981), but that distinction was too limiting, which is why the debate ultimately degenerated into synthesizing spatial metaphors, such as time as a spiral (Howe 1981). Nancy Munn's work on kula exchanges (Munn 1986) richly describes how present actions create multiple possible futures. Morten Nielsen's discussion of house building in Maputo, Mozambique, also emphasizes "continuous multiplicity" (2014, 168) rather than linearity.

Anthropology doggedly clings to its European temporal ideas—the temporalities that are essential to capitalism. Laura Bear notes that "anthropologists have used theories of time profoundly shaped by practices of modern social time" (2014a, 6), and advocates that we subject these temporalities to explicit examination. This involves attention to the epistemology of time rather than to the ontology of time (2014a, 14).

To do this, it is necessary to jettison some old baggage. The assumption that all cultures have the same conceptual foundation for time has persisted in anthropology (Nilsson 1920; Bloch 1977; Gell 1992). Nilsson argues that "at the basis" of all time reckoning "lies an accurately determined and limited and indeed small number of phenomena, which are the same for all peoples all over the globe, and can be combined only in a certain quite small number of ways" (1920, 2). The resulting assumption that cultural concepts of time can be translated into clock and Gregorian calendar rubrics without much distortion is found in many ethnographic representations.

The distortion introduced into calendars by such transformations skews modern attempts to recreate ancient practices, however. For instance, many neo-pagans attempt to recreate the festival cycle of the pre-Christian Celts, but they do so utilizing the Gregorian calendar. As a result, the holidays are not tied to the astronomical cycles that the Celts probably recognized, but to calendar dates instead. The Celtic holiday of Samhain is an example. Based on the work of the philologist Sir John Rhys (1892, 518), it is currently celebrated on November 1. James Frazer uncritically adopted Rhys' view and states that the Celtic New Year was reckoned "from the first of November" (1961, 81), as if the Julian calendar were the Celtic calendar and, consequently, the month named the "ninth month of the year" was the first. Hutton points out that Rhys does not offer conclusive evidence for this calendar date (1996, 363–364). He states, "[M]edieval records furnish no evidence that 1 November was a major pan-Celtic festival" (1996, 362).

Gell (1992) and Bloch (1977) are on firmer ground than Rhys when they argue in favor of a basic, a priori judgment of time that is true of all humans. This argument has credence, but much time reckoning is not based solely on cognitive judgments but, instead, is cognitively mediated by the environment or time-reckoning tools (Birth 2012). Time reckoning, then, is an example of what Clark (2008) calls the cognitive extension—the use of things in the world to think. To understand cultural temporalities, one must understand, then, the interaction of the mind and its environmental or artifactual extension.

This is a challenge. There is great cultural variability in the choice of such things in the world, as well as the logics behind the construction of time-keeping artifacts. In part, this is because there are so many cycles from which to choose, and they do not neatly relate to each other. Twelve lunar months do not equal a solar year; the amount of daylight varies according to the season and distance from the equator; the annual cycle of some living things is tied to temperature; the annual cycle of other living things is tied to light and time of year. If one accepts Nilsson's point that time reckoning is based on a small number of phenomena (1920, 2), that still begs the question of why certain phenomena are selected and others neglected. Why does the Gregorian calendar select the solar year, the Islamic calendar the lunar year, and the Hindu calendar seek to reconcile the differences between the solar, lunar, and sidereal years? Why did Hesiod say that one should plough when one hears the "voice of the crane" overhead as opposed to a different species of bird (1988, 50)? The cultural trend in European time reckoning has been to discard terrestrial and celestial cycles in favor of more predictable and convertible systems of counting. While those involved in developing the technologies for ever more precise measurements view this process as progress, when compared with other systems of time reckoning, it is clearly a cultural choice. The choice has been to define accuracy in terms of measurement of uniform units of duration rather than in terms of the ability to pinpoint a moment.

In European time-reckoning practices, the choices made to emphasize uniform duration as the foundation for time reckoning are not recognized as choices. The logic is naturalized and becomes foundational in how time is conceptualized. This begins at the level of acquiring data. Whether it is combing through archives or doing ethnographic research, data are organized using a temporal framework. Each bit of information is given a date, and while the process of analysis might ignore the dates, knowing the succession of events is often useful. Lederman writes that the "chronological organization is orienting" (1990, 9). Yet, while the analytic usefulness of creating a temporal grid for information is clear, the obviousness of this practice hides its artificiality.

This is not to say that there is an opposition between the artificial and the true. As I emphasized in the Prelude, all representations of time are artifices. One should not misinterpret my emphasis on the artificiality of temporal organization as an argument to discredit the practice. Such temporal organization is created for a purpose, although its use is not necessarily limited to that purpose. Likewise, the ability to time durations down to fractions of a second is also an artifice, yet it is immensely useful in determining whether records in races are broken or not and the sequence of computerized stock trades, but it is of limited utility as a feature of my watch. Following Vico's distinction between truth and certainty (Vico 1984; Berlin 1976), clock and calendar times are true in the same sense as mathematics—they are systems of representation in which the rules are well-defined. Clock and calendar time are also speech acts—created by signifying.

Since the seventeenth century in European chronologies, the artifice of time serves the purpose of creating a single temporal grid in which all events can be placed (Wilcox 1987). In this era of globalization, being able to demonstrate the synchronous occurrence of events in far-flung parts of the world is important and is part of how many people in many societies experience the world. The development of methods to demonstrate simultaneity is, in fact, a novel problem (Galison 2003)—one with which no era before ours was concerned. It is not a novel discursive phenomenon, however. For instance, in the Gospel of John (4:45–53), there is a story that asserts simultaneity without discussing any techniques for demonstrating simultaneity:

> Then he came again to Cana in Galilee where he had changed the water into wine. Now there was a royal official whose son lay ill in Capernaum. When he heard that Jesus had come from Judea to Galilee, he went and begged him to come down and heal his son, for he was at the point of death. Then Jesus said to him, "Unless you see signs and wonders you will not believe." The official said to him, "Sir, come down before my little boy dies." Jesus said to him, "Go; your son will live." The man believed the word that Jesus spoke to him and started on his way. As he was going down, his slaves met him and told him that his child was alive. *So he asked them the hour when he began to recover, and they said to him, "Yesterday at one in the afternoon the fever left him." The father realized that this was the hour when Jesus had said to him, "Your son will live."* So he himself believed, along with his whole household. Now this was the second sign that Jesus did after coming from Judea to Galilee. (New Revised Standard Version, emphasis added)

To interpret this passage, the concept of the hour used in Gospel narratives needs to be freed from its modern connotation as a unit of time. According to Isidore of Seville's *Etymologies*, the Latin word *hora*, which we translate as *hour*, was derived from the Greek word ὥρα, which Isidore describes as referring to the "edge of the sea, rivers, or clothing" (2005,

Vol. 29). So in the Gospel passage, *hour* does not refer to a block of time lasting 60 minutes, but instead refers to a significant moment, such as from a transition from sickness to health. This rhetorical assertion of simultaneity is common in Christian hagiography, such as the story of how young Saint Cuthbert saw a vision of a soul ascending to heaven the same night that Saint Aidan died (Bede 1844–1864, 740).

At some point, in the assertion of simultaneity, rhetorical representations of events happening "at the same time" gave way to rhetorical time frames that demonstrated that two events occurred within the same unit of time—what Nowotny describes as the discovery of worldwide simultaneity (1994, 22–28). This was a simultaneity different from that in Christian hagiography. It was not a simultaneity accepted by faith, but a simultaneity demonstrated through a chronological grid of dates and times linked to the emerging field of metrology. As Benedict Anderson has argued about print media (2006), global communications and media connections increase the utility of the artifice of a single temporal framework in which to place all events. The importance of the experience of simultaneity across the globe is an important component of the concept of time–space compression that has been identified as an important feature of globalization (Castells 2000; Friedman 2005; Harvey 1989, 1990, 1993). The inconsistencies of the system are rarely felt, unless one studies Jules Verne's *Around the World in Eighty Days*. We can no longer imagine Phileas Fogg miscalculating how long it took to circle the globe—today, his GPS-guided portable electronics would tell him the "true" time.

Heterochronicity

So research occurs in a context in which this uniform temporal grid is based on efforts to define time and to demonstrate simultaneity. This was not always the case. Having to relate different temporal frameworks to each other has been a common exercise in human history that we only awkwardly pursue. A case comes from Thucydides in a passage that Feeney describes as one "that no discussion of synchronism ever omits" (Feeney 2007, 17):

> The thirty years' truce that took place after the recapture of Euboea lasted for fourteen years. In the fifteenth year, when Chrysis was then in her forty-eighth year as priestess in Argos and Aenesias was ephor in Sparta and Pythodorus was archon for the Athenians with still two months to go, in

the sixth month after the battle at Potidaea and just as spring was beginning, a little over three hundred men of Thebes under the command of the Boeotarchs Pythangelus son of Phylides and Diemporus son of Onetoridies entered around the first sleep with their weapons into Plataea, a city of Boeotia that was an ally of the Athenians. (quoted in Feeney 2007, 17)

The practice of juxtaposing and relating multiple chronologies is still prevalent, but submerged unless one is part of a group that maintains its temporal traditions in the face of overwhelming secular calendrical domination. The Old Calendarists of the Eastern Orthodox Church, Muslims, Jews, Hindus, and Chinese all maintain chronologies and calendars that do not fit with the Gregorian calendar, and there are many tools for relating these different systems to one another.

Such tools are not straightforward in their algorithms. The classic treatment of this challenge is *Calendrical Calculations* by the computer scientists Nachum Dershowitz and Edward Reingold (1997). They undertook to develop algorithms to convert between the major calendar systems of the world. In the process, they came to recognize that different calendars were built on different logics. They discovered two broad logics. First, there are calendars that are based on arithmetic. In such calendars, there are units of standard length that are counted and that extend into the future without being reset. The Gregorian calendar with its leap year function is an example. The second kind of calendar they call astronomical. These calendars emphasize the calculation of astronomical events rather than the counting of units. The contrast can be grasped in the difference between the secular and the Jewish day. The former is based on 24 hours elapsing beginning at midnight. The latter is determined by twilight and sundown. The former is the same duration every day of the year. The duration of the latter varies with latitude, season, and atmospheric refraction.

Dershowitz and Reingold demonstrate that developing algorithms for the conversion between calendars is possible, but they also offer warnings that such algorithms are only temporary: "Our code will not work forever ... [T]hese results may be *culturally* wrong in the sense that, say, Copts may not refer to a year 0 or -1 ... Checking the results of conversions against the historical record is sometimes misleading because the different calendars begin their days at different times" (1997, 28–29, emphasis in original).

As a result of the global hegemony of the Gregorian calendar, the algorithms developed by Dershowitz and Reingold are only of concern to those trying to convert various calendars into a Gregorian representation. For most calendar users, non-Gregorian traditions are represented in

Gregorian calendars. It is much more difficult to find Western Christian holidays and dates represented in Muslim, Hindu, Jewish, or Chinese calendars (Birth 2012, 168). Yet these other temporalities have not disappeared. While Bhabha (1994, 139–170) describes alternative temporalities as forms of resistance, this seems to make such temporalities "marginal to modern nationhood" (Allen 2008, 9). Yet, Jewish *zmanim*, Muslim *ṣalāh*, Hindu *jyotish*, and Chinese calendrics are critical to these identities, thereby leaving those who accept Bhabha's logic with the awkward, Eurocentric conclusion that Jewish, Muslim, Hindu, and Chinese identities are marginal to modernity.

Problems in Western Chronology and Temporality

There are contexts within European time-reckoning practices in which the impossibility of uniformity creates problems. Genealogists and historians are familiar with this problem when dealing with dates before and after the change from the Julian calendar to the Gregorian calendar. In Catholic dominions, this change involved a ten-day difference in 1582; in British dominions, the change involved an 11-day difference and did not occur until 1752. In colonial dominions like the Caribbean, it is anybody's guess which calendar was used before 1752. This poses a challenge for historians and genealogists trying to make sense of calendar dates in English dominions between 1582 and 1752, and this challenge is made worse in the former colonies since references like Cheney's *Handbook of Dates for Students of English History* (Cheney 1970) are focused on the metropole.

It is not just calendar times that are confusing, but also time of day. Before the adoption of the modern system of representing time with its division of the day into 24 hours of equal length, time was reckoned by a variety of means, including dividing daylight into equal segments. This makes conversion from what a text says into modern time reckoning difficult. Take, for example, the following statement in Pliny the Elder's *Natural History*: "in Tenedo insula fons semper a tertia noctis hora in sextam ab aestivo solstitio exundat." This is translated by Rackham as "A spring on the Island of Tenedos after midsummer always overflows from 9 to 12 p.m." (1938, 356–357). The Island of Tenedos is at about 40 degrees north latitude. Around the summer solstice, there are about 15 hours of daylight on Tenedos. This means that the third hour (i.e., *tertia noctis hora*) would not be at nine o'clock, but at about 9:45.

The Gregorian calendar and clock time cannot be taken for granted, then, nor can their cultural implications and uses be assumed as the same across the globe. The taken-for-granted homochronicity that they generate hides a history of heterogeneity, conflict, and struggle (see, for instance, Allen 2008). It is worth looking at how they emerged to learn the implicit values inherent in homochronicity—a hidden logic that represents both the cultural proclivities of those who have spread Europe's distinctive time reckoning, and some of the issues against which much of the rest of the world reacts.

This leads to an inconsistency of representations of time even when dealing with cultural differences. Johannes Fabian has documented how traditional ethnographic representation was allochronic—it represented "the Other" in a time outside of the time of the ethnographer and the readers of the ethnography (2002). Yet, within this allochronic time frame, the concepts of time employed are refracted through European time concepts. In the anthropology that adopted a cultural evolutionary framework, this was reconciled by assuming that "primitive" systems of time were precursors to civilized systems. The justified removal of a unilineal evolutionary framework for analyzing cultures fostered an ethnographic present which was both timeless and included time that unfolded in Western terms—a sort of Lewis Carroll-like Wonderland in which the White Rabbit was conscious of clock time passing but it was always teatime for the Mad Hatter. This is the sort of allochronism Fabian identifies (2002). But removing the sense of being in another time or timelessness upon which ethnographic description used to rely did not completely solve the dilemmas of time facing anthropology. There remained the imposition of Western time concepts on the description of non-Western other rhythms and time systems.

For some ethnographers, like myself, clock and calendar times were features of the social landscape. Every home in Trinidad seemingly has clocks and a calendar. In fact, calendars are a favored marketing device of small businesses—they print calendars with their business's name and distribute them to customers. It would be easy to represent clock and calendar times in Trinidad as nothing unusual, unless one wishes to take the position (which is where I am leading) that clock and Gregorian calendar times are unusual wherever they are found. Yet, there are indications that the assumption of temporal coherence fostered by European-style time reckoning is a false one, even within European-derived systems.

Even the dominant chronology of the BC/AD distinction is off from its reference point. The use of this distinction came into common use in the seventeenth century, and even though some have changed the terms from "Before Christ" and "Anno Domini" to "Before the Common Era" or BCE and "Common Era" or CE, the anchor point of the chronology is still supposedly the birth of Jesus Christ.

The year of Jesus' birth that is used for this chronology was calculated by Dionysius Exiguus in a seemingly rather matter-of-fact way:

> Si nosse vis quotus sit annus ab incarnatione Domini nostri Jesu Christi, computa quindecies XXXIV, fiunt DX; iis semper adde XII regulares, siunt DXXII; adde etiam indictionem anni cujus volueris ... Isti sunt anni ab incarnatione Domini.
>
> [If you wish to calculate the year of the incarnation of our Lord Jesus Christ, calculate 15 times 34, which makes 510; to this always add 12, which makes 522; add also the year of your current indiction cycle [a 15-year cycle] ... These are the years since the incarnation of the Lord]. (Dionysius Exiguus 1844–1864, 499)

Even at the time that Dionysius Exiguus did his calculations, there were competing models (see Blackburn and Holford-Stevens 1999, 773; Declerq 2000, 114), and current interpretations suggest that his calculation was off by four to six years.

So European-derived clocks, the Gregorian calendar, and chronologies all do not reflect their original temporal anchors. As mentioned in the previous chapter, clocks are no longer derived from the apparent day. The Gregorian calendar does not accurately reflect the solar year, and the BC/AD (or BCE/CE, if one prefers) chronology is set to a likely inaccurate estimation of the year of Jesus' birth. This separation of systems of time from their original cues is not unusual. Nilsson wrote, "In the more fully developed calendars there are not seldom found periods of time which are reckoned without reference to any of the factors given by nature" (1920, 324). The rejection of cycles "given by nature" is because of their irregularity. European timescales consist of homogeneous units treated as if they are of equal duration. European timescales emphasize homogeneity and long-term stability over observation and correction. All of these assumptions get absorbed into and subtly structure the representation of time in pre-Enlightenment and non-Western societies.

Hegemonic Calibration

Even though European-derived timescales' ontological foundations are shaky, they have become the dominant temporal standard for the world, a standard to which all other times must be calibrated. The Gregorian calendar and European chronologies are used to chart all events and all calendrical systems.

Homogeneous time involves a subtle imposition of power in the form of clock time. In its current form, it diverts attention away from individually observed environmental cues to times that are centrally calculated and distributed. Clocks, calendars, and chronologies homogenize time, and their development has witnessed an important ontological shift away from its being constituted in relationship to the Sun and toward its being determined by a statistical representation of the periods of atoms in laboratories distributed throughout the world. This change in the determination of chronometrics indicates its homogenizing tendency, in that Earth's rotation was too erratic for modern science and Christian hegemony too powerful to reform the calendar and chronology any further.

Homogenous clock time, then, is an approximation of a concept of time divorced from existence. Yet, it is viewed as part of nature. This denatured natural time is a symptom of the post-Enlightenment transformation of nature away from that which is merely observed to that which is manipulated—a view of nature "reconceived as manipulatable material, determinate, homogeneous, and subject to mechanical laws" (Asad 2003, 27). Time reckoning involved a literal movement from nature to mechanics—about eighteenth-century England, Berthoud writes, "We are, in fact, witnessing the birth of a new mentality. Formerly, from public time—the sun in the sky and the clock in the tower—men had learnt time-obedience; the invention of private time—the indoor clock and the pocket watch—is now teaching them something new: time-discipline" (1987, 39–40).

It is a "truth" that humans have created and that many unquestioningly accept even when trying to understand temporalities based on very different assumptions. The physical sciences have no compunction about using the clock to measure time. Even studies of circadian rhythms that are triggered by sunlight use the time indicated by clocks that are set to UTC as the standard of measuring the rhythms. There is not even any acknowledgment of the irregularity of Earth's rotation, and consequently of sunlight—one anonymous reviewer of my article "Time and the Biological Consequences of Globalization" (Birth 2007) stated that this deviation is too insignificant to be of interest. This might be true, but chronobiology also rarely controls for time zones or latitude. In a time zone, all clocks are set to a single longitude within the zone, and consequently can deviate from the local solar time by four minutes per degree longitude, or, ironically, four temporal seconds for each minute of longitude. (A riddle: When does a minute equal four seconds?) Latitude is even more of a problem when thinking about daylight. The seasonal variations of daylight increase with higher latitudes. Circadian cycles triggered by sunlight at the equator are far more stable throughout the year than similar circadian cycles near the Arctic or Antarctic. Horological homochronicity has been imposed on our understanding of processes in nature.

CHAPTER 2

Evolution's Anticipation of Horology?

Henzi et al. write that the baboons in the Drakensberg Mountains had a period of constant movement that averaged "from 0900 to 1700 hours" (1992, 613). In effect, the baboons worked from 9:00 a.m. to 5:00 p.m. Put this way, the clock time seems to create a parallel between nonhuman primates and human laborers working an eight-hour day. Instead of reading any evolutionary significance into the baboons' eight-hour workday, one should see it as a symptom of the primatologists' unreflective use of clock time. To my knowledge, baboons are not clock watchers, but have cycles of activity tied to cycles of daylight, and daylight is seasonally variable. So describing their activity cycles in terms of an "average" and using clock time hides the phase relationship between sunrise and activity. In so doing, the primatologists unwittingly obscure an important feature of baboons' behavior in relationship to their ecological temporal niche.

When one thinks about it, it is astounding that a cultural artifact invented in Europe in the thirteenth century now dominates how we think about all timing processes, including those produced by evolutionary processes. It is as if evolution anticipated a humanly created device by millions of years. Because clocks are important cognitive tools for the measurement of duration, they often subtly mediate between timing phenomena and the modeling of those phenomena. The result is that often the tool becomes confused with the data. When this happens, the logics embedded in the tool can warp understanding. Such is the case with many studies of circadian cycles.

The social and cultural dimensions of scientific practice are now well documented (Galison 1987; Kuhn 1970; Latour and Woolgar 1986; Traweek 1988). Less attention has been given to how the tools used in scientific research are cultural artifacts that embody culturally created algorithms for solving particular cognitive problems. In the case of concepts of time, as Barbara Adam points out:

> Newtonian science recognizes no contextually based differences in rhythm and intensity, no contextual tempo or timing, duration or change, no times inherent in processes and phenomena, no force that constitutes *natura naturans*, no *Wirkwelt* that works below the surface, no life and death, growth and decay, no seasonality, no right time for every season and place, no special days and moments or difference between sacred and profane times, no stress and pressure of "deadlines", no decorum, no valorization of speed, no reverence for the past, no hopes and features for the future. (1998, 40)

It is Newton's idea of abstract, absolute time that dominates the measurement of time and, consequently, laboratory science. This can become an obstacle to applying what is learned in experimental, laboratory settings to the study of timing and rhythms outside the laboratory. Cycles and rhythms "in the wild" are not governed by the clock—unless clock time is imposed.

Homogeneous uniform time, then, can obscure nonuniform, but still rhythmic or cyclical, biological processes. For instance, in his study of hospitals, Zerubavel noted, "Though nature recognizes no regular hourly rhythmicity, all medications are officially supposed to be administered on the hour" (1979, 30). The body's metabolic processes vary considerably throughout the day, and this has been shown to have an effect on the effectiveness of many medications (Dallmann et al. 2014), yet the dominant model for the use of drugs is that the time of day does not matter as much as the duration between doses. Adam also notes how calendar and clock times are imposed on pregnancy and giving birth (1995, 48–52). This imposition elides the relationship between cultural behavior and biological processes. Time is not outside of nature. On the contrary, as Elias points out, "If one explores 'time' one explores people within nature, not people and nature set apart" (1992, 97). Nature is not a constant, nor is it immune to human activity. To understand time in terms of people in relationship to nature is to recognize the relationship as a dynamic one.

Chronobiology

Today, Trinidad and Tobago is malaria-free—an example of humans shaping their environment in a profoundly important way. A by-product of this achievement was a set of revolutionary insights into chronobiology—the study of biological cycles and rhythms.

In the 1940s, a young Colin Pittendrigh arrived in rural Trinidad. He had been hired by the Rockefeller Foundation to conduct research on malaria-bearing mosquitos in order to control, if not to eliminate, malaria from Trinidad. Pittendrigh would later go on to become a pioneer in chronobiology, but when he first arrived as a young man in Trinidad, chronobiology was not a field of study.

Among the places Pittendrigh studied mosquitos, he was particularly interested in a community I call Anamat—the same Trinidadian village where I conducted ethnographic field research. In Trinidad, there were two species of mosquito that carried malaria: *Anopheles bellator* and *Anopheles homunculus*. In most of the island, their ranges did not overlap, but in Anamat they did. This fact drew Pittendrigh to this remote location "in the bush."

He was given permission to live in the house on the grounds of the local Department of Public Works. This house was a short walk from the cemetery—a small rise in the middle of which stood a tall tree. Dr. Pitt, as he was known locally (even though he had not received a doctorate yet), hired boys to climb this tree and drop mosquito nets at timed intervals on a donkey tethered to the tree. By counting the mosquitos that had alighted on the donkey, Pittendrigh recognized that the malaria-carrying species of mosquitos were not active at the same time, and this insight led him to start to speculate about the idea of a temporal niche that complemented the idea of the ecological niche, as well as biologically determined rhythms of behavior (1950, 1954). In the case of mosquitos, it did not seem to be a particular duration after sunset that prompted their biting, but humidity. *Anopheles bellator*'s and *Anopheles homunculus*'s biting behavior was triggered by different humidity levels with the result of their engaging in their blood sucking at different times. This research drove Pittendrigh's doctoral work. For the elderly men that I interviewed, their work for Pittendrigh was their first memory of a link between work-discipline and labor-time.

What Pittendrigh's early work demonstrated is a point I made in the last chapter. Nilsson claimed that cultural ideas of time are derived from environmental cycles, but I suggested that there are so many environmental cycles from which to choose that Nilsson's idea begged the question of the cultural logics behind the choices. Mosquitos are part of the environment to which cultural practices respond. The daily rhythm of life in rural Trinidad includes an awareness of when mosquitos will start to bite, at which time many people light mosquito coils or plug in "bug mats" to keep the pests at bay. This does not mean that there is a cultural marking of "mosquito time," merely that the biting behavior of mosquitos is among the many environmental cycles that shape cultural practice.

Mosquitos are just two of the organisms living in Trinidad, and each has its own cycle. Some cycles, like those of mosquitos, are tied to daily changes in humidity levels; others, like the changing toughness of bull grass, which will be discussed in the next chapter, seem related to sunlight and temperature.

Despite Pittendrigh's early work that emphasized different rhythms of different species of insect, the early literature on biological cycles was already mired in the metaphor of the clock. For instance, Maynard Johnson described mice as having "an exceptionally substantial and durable self-winding and self-regulating physiological clock, the mechanism of which remains to be worked out" (1939, 326). By the Symposium on Biological Clocks held at Cold Spring Harbor in 1960, a seminal meeting in the development of chronobiology, the clock metaphor was widely used, yet, as was the case with research on psychological timing, the metaphor was based more on how ordinary clock users approach clocks rather than on the technical aspects of how clocks kept time. For instance, in his introductory remarks for the symposium, Bünning, another pioneer in chronobiology, stated, "We have, figuratively speaking, damaged the hands of the clock or violently turned them with our fingers, but we have not influenced the clockwork itself" (1960, 7). In reading Bünning and others, it becomes clear that they know how to use clocks, but not how clocks work. In a sleight of hand, the metaphor based on clock use became the metaphor for how biological clocks functioned.

The fascination with the setting of the clock's hands was an outcome of the growing concern with entrainment in the 1950s. Entrainment involves the environmental factors that "set the clock." Despite Pittendrigh's research that emphasized humidity, as chronobiology developed, research questions were often based on factors that influence the performance of mechanical clocks and watches such as temperature and electromagnetism. The chronobiological research questions were driven by the knowledge of clock consumers, not clockmakers. Technical horological knowledge was not the source of metaphor, and the actual time-keeping mechanisms of clocks and watches (gears, escapements, balances, springs, etc.) did not influence chronobiology. These clock-derived questions eventually led to conclusions that biological timing was not subject to the same influences as mechanical devices.

Pittendrigh himself admitted that, in the early period of research, "the phenomenon of temperature-compensation seemed the outstanding problem" (1960, 179). As a result, early chronobiology went through a phase of studying the influence of temperature on circadian clocks only to determine that temperature changes were not significant. Pittendrigh did a series of experiments that disproved Bünning's thesis that temperature influenced the tempo of biological cycles, yet Pittendrigh continued to use the metaphor of clocks. Based on these experiments, he argued, "It is a logical necessity that the fly must possess, as part of its regulatory equipment, the equivalent of a temperature-insensitive time-piece" (1954, 1027), and that this clock be "a piece of inherited physiological machinery" (1954, 1025). Pittendrigh would later state that circadian cycles were driven by oscillations that were "sufficiently temperature-compensated to serve as a useful clock" (1993, 26).

With regard to electromagnetism, Rütger Wever attempted to show its influence on the biological cycles. He did a series of experiments from which he concluded that a 10 Hz electrical field "shortens the autonomous period, diminishes the interindividual variability, and reduces the tendency toward internal desynchronization" (1979, 111). He then added that the results from the room shielded from natural electromagnetism with a 10 Hz field introduced produced the same results as the non-shielded room. Rather than conclude that an electric field had no influence on internal timing, he concluded that 10 Hz matched environmental electromagnetism. Later evaluations of the effect of electromagnetic fields on circadian rhythms have also produced equivocal results (Paneth 1993; Warman et al. 2003).

In effect, with the subtle focus on whether factors that influence mechanical clock performance influenced biological "clock" performance, research moved away from addressing the concept of the temporal niche and the competitive or complementary relationships of the cycles of different organisms that shared the same environment—including humans, whose activities are influential on mosquitos (see Livingstone 1958). Chronobiology could explore human-animal interaction, but it does not explicitly do so. It only implicitly addresses this through the human-imposed controls of laboratory experimentation with animals. In such human-imposed conditions, it is clock time that structures the experimental animals' environment. To some extent, much of chronobiology is the study of how well animals can adapt to humanly created times. As Nowotny puts it, "[N]ature is not just observed, but reproduced and generated in the laboratory ... [T]he social construction of the natural world begins" (1994, 88).

To be fair, even though ideas of clock performance shaped research questions about biological cycles, there was a recognition that there were significant differences between the biological clock and mechanical clocks. For instance, Aschoff wrote, "Watches are instruments to measure time. Organisms have to measure time-spans of quite different lengths and for different purposes ... One can expect, therefore, the organisms possess several clocks with perhaps extremely different periods" (1960, 11). Pittendrigh stated it strongly: "We are forced, in fact, to abandon the common current view that our problem is to isolate and analyze '*the* endogenous rhythm,' or '*the* internal clock,' and are faced with the conclusion that the organism comprises a population of quasi-autonomous oscillatory systems" (1960, 165). This idea of several "clocks" supported the theory of multiple oscillators guiding circadian patterns. In this multi-oscillator model, a crucial component was how the oscillators were coordinated. As this model developed, more and more attention was being focused on the free-running behavior of the oscillators. In other words, if the clock was deprived of external cues, what sort of time would it keep?

The reason for this lay in the methodological differences between laboratory research and field research. A lot of early chronobiological research still had a field component that measured multiple cycles. This methodological difference cultivated relationships to clocks as a research tool that differed from laboratory-based research. In field studies, the clock was merely a tool for measuring environmental cycles and not a means of controlling variables. In the laboratory, clocks are used to constitute environmental cycles as well as to measure timing variables.

While chronobiology struggled to keep the clock metaphor and biological timing conceptually separate, the move from field observation to laboratory studies resulted in more and more variables, including timing variables, being controlled by the clock rather than by features of the observed environment. The use of clocks to measure phenomena and the use of clocks to control phenomena risked becoming conflated.

To conform to Newtonian logic of uniform timekeeping, laboratory environments often constructed stable cycles of light, temperature, or food availability that were unlike conditions animals (including humans) faced outside the laboratory.

An example of this is how light cycles were controlled. Laboratory studies of biological cycles control light/dark cycles. As a result, the prevalence of 12-hour cycles of light and dark in the laboratory has unwittingly generated knowledge about how chronobiology functions in the tropics on sunny days, but not about how the majority of species on this planet which are not in the tropics, and which must contend with a variety of weather conditions, function. For instance, the difference between the light/dark cycles of many laboratory mice and wild mice can be easily seen in Figure 2.1. Imagine that the mouse in question lives in, say, northern England where it is often cloudy, and the cloud cover creates daily fluctuations in the duration and intensity of light. The curve representing the hours of light at 60 degrees north latitude is then too neat a picture.

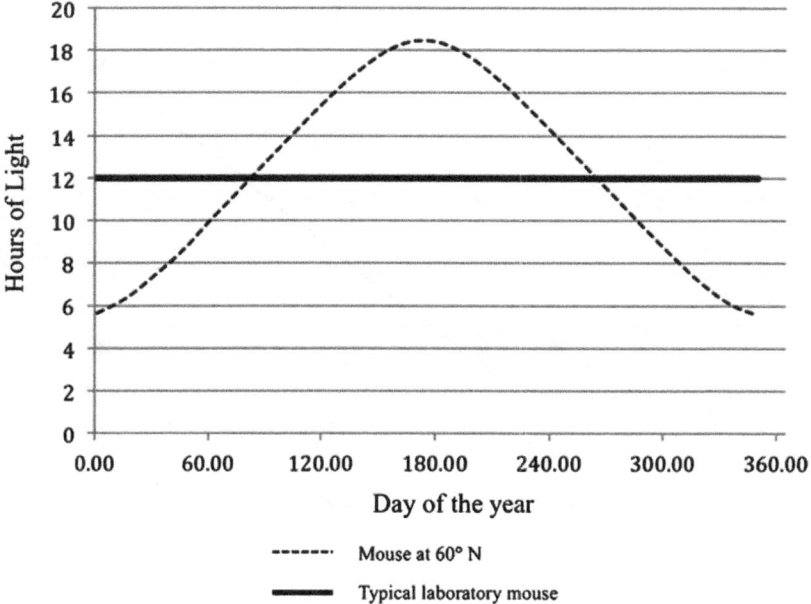

Fig. 2.1 Light/dark cycles of mice at 60° N versus laboratory mice

In addition, how light is introduced into laboratory conditions is quite different from how it is experienced outside the laboratory. Figure 2.2 displays the change of light at the beginning of the day in the laboratory versus in the wild. In the world, light gradually increases from less than 1 lux to 40 lux at sunrise, to 1000 lux one hour after sunrise, to 10,000 lux at midday. In the laboratory, light typically increases from less than 1 lux to a maximum of around 500 lux instantly.

The use of environments that are temporally constructed using clock time shapes research results. On the one hand, they attest to the potential of the biological rhythms to adapt to humanly created conditions, but they only give a murky glimpse into the evolutionary significance of the biological cycles. The murkiness is generated by the use of artificial light and clocks. As a result, the environment for all organisms in which chronobiology has been studied is remarkably similar in its cycles to that of the equator even though the majority of species in which biological cycles have been studied have ranges not limited to the tropics.

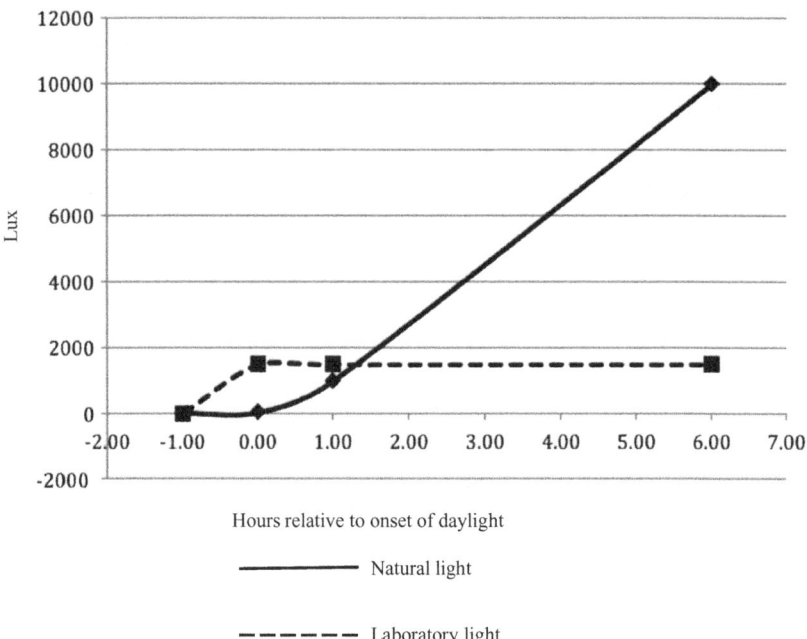

Fig. 2.2 Differences in light between a sunny day and a laboratory

As the focus shifted toward studying the molecular and genetic components of free-running cycles, and as in the study of psychological timing (see Birth 2014a), the emphasis of the metaphor shifted away from how the system was like mechanical clocks to the synchronization of clocks across networks. For instance, in a recent article entitled "Central Control of Peripheral Circadian Oscillators," Menaker et al. (2013) address the issue of how multiple oscillators distributed throughout the body are synchronized. This article challenged the idea of a centralized pacemaker in the brain in favor of a model of "a network in which signals from a primary central pacemaker regulate the phases of second-order pacemakers, which in turn control subsets of peripheral oscillators, some of which may in addition be tertiary pacemakers. The inevitability of feedback in such a system makes it into a network" (2013, 741).

This model might or might not be true of biology, but it is an absolutely correct representation of how time synchronization occurs on computer and telecommunications networks. Most such networks rely on Network Time Protocol (NTP). In these networks, individual devices have their own clocks that can run independently of any synchronization, but tend to be poor-performing timekeepers. For these clocks to be synchronized to a time standard, they send and receive time signals from more reliable clocks in a hierarchical arrangement. The highest-performing clocks, say, in national time laboratories, are known as stratum 0 time sources. Stratum 0 sources synchronize stratum 1 sources, and stratum 1 sources synchronize stratum 2 sources all the way down to one's personal electronic devices (Mills 2011). In effect, even though current models of biological timing no longer frame their questions in terms of the performance of mechanical clocks, they still seem to mimic how current computer time standards are maintained and disseminated.

As this research documented the existence of the "clock," neuroscience has sought to isolate where the clock functions were located, and in this regard, it has been much more successful than those working on interval timing. In mammals, the crucial neural structure associated with timing and entrainment is the suprachiasmatic nucleus (SCN). At the base of the hypothalamus is a small bundle of neurons called the SCN. This structure coordinates the timing of biological processes in mammals. It consists of two parts. There is a ventrolateral area that is weakly rhythmic and retinoresponsive. The function of this set of neurons is to receive light signals to entrain the other part of the SCN and consequently to entrain the body to local patterns of daylight. The ventrolateral area connects to a dorsomedial region that is rhythmic and nonretinoresponsive. Instead, this region of the SCN has a stable rhythm of around 24 hours. It is not exactly 24 hours, however, because cycles of daylight are not exactly 24 hours in length. The function of this area is to maintain a cycle that can free run throughout the day and be reset the next sunrise. Together, the two components of the SCN adapt the organism to local variations in the timing of daylight based on seasons and weather. As a result, seasonal variations typically do not cause any biological trouble and cloudy weather does not upset circadian cycles.

Research on a single timing mechanism runs the risk of confusing the clock as a research tool with the clock as a metaphor for what was being studied. Those who examine the free-running characteristics of the dorsomedial region of the SCN without noting that the SCN is a mechanism that entrains endogenous rhythms to light cycles tend to be laboratory based and to use controls to focus on a single element of timing. Not surprisingly, they often exhibit confusion of clock and biological phenomena. The free-running cycle of humans is not a good timing mechanism without entrainment by light, as the day/night reversal relative to clock time of some Eskimos during the winter season demonstrates (Stern 2003). Rather than mimicking clocks and uniform timing, the SCN evolved to match circadian rhythms with the foibles of a tilted, wobbling planet moving on an elliptical orbit with a slightly irregular rotation.

The SCN's importance to circadian cycles is well documented, and now the major thrust of research is to understand its working at the molecular level. The SCN is a part of the brain that is described as highly responsive to cycles of light and dark, and a part that plays a crucial role in triggering many of the body's daily cycles. Often, it is described as environmentally sensitive. Yet, the science of the SCN seems to have gone slightly awry by deflecting attention from what it does toward an image of a de-globed SCN. To demonstrate this, one can simply do a search for "SCN" in the *Journal of Biological Rhythms* and look at how the SCN is discussed. On February 23, 2015, I did such a search and organized the results by displaying the most recent first. I focused on those articles that dealt solely with the dorsomedial portion of the SCN—the most common way the SCN is studied and the most likely to ignore the SCN as a means of adapting to the environment. Of the first five results, two emphasize the clock metaphor in the title. The first such article was "Adrenal Clocks and the Role of Adrenal Hormones in the Regulation of Circadian Physiology" (Leliavski et al. 2015). The first sentence of the abstract was "The mammalian circadian timing system consists of a master pacemaker in the suprachiasmatic nucleus (SCN) and subordinate clocks" (2015, 20)—again, the same structure and organization as NTP. The second sentence of the article was "Circadian clocks (from the Latin *circa diem*—about a day) allow for reliable timekeeping with a period of close to 24 h without external stimuli but are synchronized to geophysical time by signals such as light and food" (2015, 20). The second article I found was "Circadian Clock Function in the Mammalian Ovary" (Sellix 2015). This article argues that "Each cell type of the ovary, including theca cells, granulosa cells, and oocytes, harbor a molecular clock" (2015, 7). A third, "Phase Resetting in Duper Hamsters" (Manoogian et al. 2015), studies the "molecular clock mechanism" (2015, 14). The two other studies did not emphasize the clock metaphor. One focused on the sensitivity of nerve cells in the retina to light and how it affects the SCN (Gompf et al. 2015), and the other on seasonal adaptations to differing day lengths (Jarjisian et al. 2015). In effect, my suggestion that laboratory-based research that focuses on endogenous rhythms tends to emphasize clocks, whereas research on adaptive fitness does not, holds for four of the five articles.

So while there seems to be nothing controversial about postulating a biological clock, this metaphor tends to lead to research that neglects the relationship of circadian rhythms and adaptation to a changing environment.

But if the SCN mediates between the organism and the environment, and the species studies are not being studied in the environmental niches in which they naturally live, what do we really know about the SCN? Because of the tacit acceptance of homogeneous time periods of light and darkness, the SCN has come to be regarded as a stabilizing feature in circadian rhythms. We do not understand its flexibility even though we know of animals that must be temporally flexible in regard to light.

A consequence is that we do not understand the relationship between temporal niche and location on the globe—in effect, we treat the world as if it were flat (Birth 2008). And with regard to humans, we do not have a link between the consequences of our expanding beyond our temporal niches and the neuroendocrine mechanisms underlying these consequences. For instance, we know the negative health consequences of shift work, but little of the biological processes that mediate between this behavior and its health effects. We also know about seasonal affective disorder and have ethnographic evidence that circumpolar indigenous peoples suffer from it (Condon 1983), but have not thought about whether this is a product of our choosing to live outside our temporal niche.

This raises a more general point about laboratory work, and the need for it to be coupled with field research. Often, what gets controlled in the laboratory might be an important environmental variable. For some functions, this is probably not an issue, but for any aspect of the body that is designed to respond to environmental features that are not stable, this is a huge conceptual problem.

The problem is not with using metaphor. Instead, the issue is the role of the clock metaphor in research. There is a danger that we may come to think that biology works like the model users have of clocks, as opposed to the complex models of time metrologists or horologists. This leads to confusion of the clock as a measurement tool with a metaphor of the clock shaping our understanding of circadian biology despite biology's many non-clocklike features. Biological mechanisms did not evolve in response to the clock, or even to clock time, and a more accurate picture of these mechanisms will emerge when they are studied in relationship to the pressures that acted upon their selection rather than in relationship to homogeneous, uniform clock time.

Conclusion

Barbara Adam writes, "Unlike the variable rhythms of nature, the invariant, precise measurement is a human invention and in our society it is this created time which has become dominant to the extent that it is related to as time *per se*, as if there were no other time" (1995, 25). Hassan argues that clock time has "displaced and dulled our sensitivity to other temporalities that exist in the natural world and in our very bodies" (2003, 27). Is the clock a good representation for biological timing processes, then?

The controlled light/dark cycles of the laboratory are unlike the conditions in which the SCN evolved. More important is how methodological decisions result in the unwarranted dominance of the clock metaphor. When single variables are studied in the laboratory using clock measurements, there is a strong tendency to treat biological systems as if they functioned like clocks; when multiple variables are studied or adaptive fitness is emphasized, the flexibility of the biological systems results in a minimizing of the clocklike metaphor, if not its absolute absence.

The responsiveness of internal timing and the SCN to environmental influences is also not clocklike. Whereas the history of clocks has been a story of innovations to limit environmental influences on clock performance, biological systems should not be viewed in this way—it makes no evolutionary sense to suggest that timing mechanisms are not part of an organism's ability to respond to variable environments. The evidence for environmental influences on internal timing and the SCN supports the evolutionary importance of internal timing. The only clear similarity between internal timing mechanisms, the SCN, and clocks is that they all indicate the passage of time, but that resemblance is insufficient to justify adopting the clock metaphor as the basis of a model for internal timing. Indeed, the clock metaphor seems to have been adopted because clocks were the most common means of thinking about time in the twentieth century. As such, they became a cultural lens through which all timing behavior was viewed. The logic behind this lens privileges units of duration over auspicious moments—of homogeneous time versus optimal timing.

In Trinidad, where mosquitos prompted the development of chronobiology, the interactions between humans and their environment are not sufficiently represented by homogeneous, uniform time. In rural Trinidad, many environmental cycles influence work. Not only does chronobiology ignore this complexity, but it unwittingly colludes with labor economics to suppress how such environmental complexity influences work patterns. A great deal of work that involves plants and animals is shaped by the circadian cycles of those plants and animals, and what drives those cycles are not clock-defined schedules. Yet, such clock-defined schedules are what organize not only laboratory understandings of biology, but management approaches to human labor.

CHAPTER 3

"Hours Don't Make Work": Kairos, Chronos, and the Spirit of Work in Trinidad

Marx wrote, "Just as motion is measured by time, so is labour by *labour-time* ... Labour-time is measured in terms of the natural units of time, *i.e.*, hours, days, weeks, etc." (1970, 30, emphasis in original). Marx's "natural units of time" are cultural creations. Even the idea of the day varies according to when it begins and ends, and whether it is measured in reference to mean time or reckoned in relationship to daily observations of solar cycles. While Marx is normally not identified with theories of scientific management, such as that of Frederick Winslow Taylor, the cultural logic that undergirds Marx's idea of labor-time is the same as that which undergirds Taylor's scientific management and time studies. In a way, both Marx and Taylor reflect the transformation from a task orientation to a time orientation in representing labor that E.P. Thompson documents in his article "Time, Work-Discipline, and Industrial Capitalism" (1967).

Thompson's historical account has been shown to be flawed, however. It places the widespread availability of clock time too late (see Glennie and Thrift 2009) and it sees the clock as a necessary condition for tying wages to time, when such practices predate the use of the clock (see Birth 2012, 62–69). Yet, Thompson does represent the perspective on the relationship of time and work that comes to dominate how economists, scientific managers, historians, and others represent labor—that units of labor are best conceptualized as units of time measured by clocks.

The conceptual problem of labor-time is how to define it. Marx struggled with this in his *A Contribution to a Critique of Political Economy*. He wrote, "To measure the exchange-value of commodities by the labour-time they contain, the different kinds of labour have to be reduced to uniform, homogeneous, simple labour, in short to labour of uniform quality, whose only difference, therefore, is quantity" (1970, 30). This is no easy task: Is this uniform labor represented by the average of one worker over time, or the average of multiple workers? Is it an average of averages? Or, is it defined by some other means? Marx elides this problem by creating the image of the "average person" whose simple labor is "average labor" with the admission that different societies and historical epochs would have different averages (1970, 31).

In volume 1 of *Capital*, he avoids this tangle of issues. Indeed, it would be scientific management that provided the methodology for defining the slippery concept of average labor-time. According to Taylor, this involved "the scientific selection of the workman" (1911, 43) who would be used to determine the standard for all workers. Taylor approached this problem as an engineer—seeing the worker as an extension of the machine (Kanigel 1997). Consequently, his selection of a worker for the presentation of all work reflected a preference for workers who were competent and efficient extensions/users of the tools and machines of their trade. Others developed their own form of scientific management that was critical of Taylor's approach. For instance, Edward Cadbury disagreed with Taylor's "reduction of the workman to a living tool" (1914, 105). The alternative proposed by British Quaker chocolate manufacturers, such as Cadbury's and Rowntree's, was to have the average worker jointly defined by management and employees (see Fitzgerald 1988). In effect, the conceptual problem of how to define average labor-time which Marx elides became a point of contention within the field of scientific management, but the contention was over how a representative worker would be selected, not over how work would be measured. All agreed that the unit of measure would be clock time.

Marx, Taylor, Edward Cadbury, and Joseph and Seebohm Rowntree were not ideologically similar, but they were all shaped by the culture of timekeeping and measurement that surrounded them—the same cultural logic that led to Derrida's misreading of Baudelaire, early ethnographers imposing clock and calendar time on ethnographic representations, and primatologists suggesting that baboons work from 9:00 a.m. until 5:00 p.m.

In contrast to this cultural logic, as I argued in *Objects of Time*, medieval European representations of time did not consistently conflate duration and timing. Moments could be as readily defined by intersecting cycles as by measured duration—the contrast between representing baboons' activity according to clock time versus representing it as a phase relationship to seasonal variations in the timing of sunrise and the amount of light. It is only through the clock and the emergence of mean time in Europe (time derived from an average of the duration of Earth's days during a year) that duration became habitually privileged over timing—first in Europe and its colonies, and eventually globally. The conflation of timing and duration in the representation of time elides the problem that timing cannot always be determined by duration. Indeed, in the United States there is an annual ritual to demonstrate this: the Thanksgiving dinner. The traditional main course for this feast is a roasted turkey. Many cooks roast only one turkey a year, and due to their lack of experience, they follow the directions that come with the raw turkey—directions that typically indicate the amount of time per pound that the turkey should be roasted. If one follows the instructions based on time, one produces either an undercooked turkey, or an overcooked and dried-out turkey. Experienced turkey cooks know that the time given in the instructions is only a rough estimate and that one must use a meat thermometer to determine when the turkey is really done. Sometimes it is done before the estimate, sometimes after, because, as the instructions always state, "cooking times may vary."

So determining when to serve a turkey based on duration is not conducive to serving the turkey at the best time. This yearly challenge in American households captures the slippage between timing and duration. Keeping these two facets of time separate is analytically useful. Despite the privileging of duration over timing in the representation of time, many processes cannot be represented in terms of homogeneous durations. Cooking times do vary.

Such an emphasis on timing has been dubbed *kairos* after the Greek deity of rhetoric. Kairos has probably received the most attention in two scholarly domains: rhetoric and theology. In rhetoric, it can be traced to Aristotle's *The Art of Rhetoric* (Aristotle 1926, 1.1.7.1354b3–1354b8; Kinneavy 2002) and Plato's *Phaedrus* (1914, 553 [272]). It should be emphasized that for Aristotle, rhetoric is not simply the art of persuasion, but the work of discovering what is persuasive (Book 1, Chaps. 1 and 2). Kinneavy and Eskin (1994, 133) point out that Aristotle's approach to persuasiveness included the qualities of individual situations. Therefore, rhetoric is not merely a set of general techniques and principles, but the understanding of the moment in which one wishes to persuade.

For Aristotle, kairos had a political component (Kinneavy 2002)—in his concepts of justice and equity, he emphasized understanding the individual circumstances of specific cases (Kinneavy and Eskin 1994). As one maxim about the law puts it, "Justice delayed is justice denied." Examples of this abound, such as the story of the suspected Nazi war criminal László Csatáry finally being located in Budapest in 2012. He was 97 years old at the time of his arrest, and the arrest took place over 67 years after the end of World War II in Europe. Given the horrors of the death camps to which Csatáry deported Jews, whatever punishment he may receive if convicted hardly feels just under these circumstances.

Kairos is an alternative to conceptualizing time in terms of duration. It emphasizes proper timing over duration, yet economics of labor emphasize duration and not timing. In contrast, practice in modern markets, such as in the context of high-frequency trading, emphasizes interventions at the right moment—in effect, kairos. This chapter explores the tension between assumptions of temporal uniformity and the kairotic experience of labor as experienced by Trinidadian workers.

Homogenous Duration and Economic Theory

In much of economic theory about capitalism, time is conceptualized as consisting of containers to be filled. In this metaphor, containers are of a specific, uniform duration, such as the hour. A common artifact of this metaphor is the datebook, a graphic representation of time as a set of boxes to be filled with commitments, but there are many other examples, such as work schedules and account books. In the metaphor of time as a container, time not only consists of units to be filled, but these units can be used to measure any activity. This second idea is particularly applied to how labor is conceived. Classical economics' concept of average labor-time represented in terms of duration is an example.

In classical economics, average labor-time is viewed as contributing to the value of commodities; in neoclassical economics, it is a means of representing the supply of labor in terms of *person-hours* (formerly *man-hours*), which, in relationship to the demand for labor, influences the price of labor. The foundation for the capitalist privileging of homogenous duration preexisted capitalism and was part of temporal sensibilities that could be found as early as the Roman Empire. The Roman army employed uniform durations to manage its activities. The standard marching pace involved a set distance in a set time (Vegetius 1967, 1.9), and presumably that pace did not vary as a unit marched northward and the period between solar hours changed with the latitude. In another example, in training his soldiers, Scipio is described as fixing the time within which building a fortified camp should be finished (Appian 1912, 6.14.86). Uniform time units were an important component of creating and maintaining army discipline—of forging uniformity among a diverse population of soldiers. Capitalism did not create uniform time units. Instead, capitalist management gradually adopted and privileged such units, because of the disciplinary and bookkeeping advantages of imposing uniformity on the diversity of workers, seasonally variable activities, and differences in daylight (and consequently the workday) by latitude.

Time Representing Labor

The contrast of duration versus timing is important for understanding the consequences of the emphasis on units of uniform duration in economic theory and capitalism. The concept of average labor-time homogenizes durations and suppresses the importance of timing. As a result, within this logic it is difficult to accept the possibility that as much can be accomplished in six hours as in eight hours, even though that was demonstrated in Kellogg's experiment with the six-hour workday (Hunnicutt 1996). Moreover, there is even more resistance to the idea that if six hours are as productive as eight hours, the workers working those shorter hours should be paid the same amount as if they worked the longer hours. The cases of disorientation and disagreement about work that I found in Trinidad are the result of disjuncture between concepts of duration and concepts of timing. In effect, the alienation of workers from their labor is not an alienation from work, but an alienation from an experience of work as meaningful and personal that is grounded in an experience of time at odds with commodified average labor-time.

The Caribbean has been an important site for such alienation and the emergence of economic theories that foster it. Sir W. Arthur Lewis, a Nobel prize–winning economist from St. Lucia, created a theory of development in which the theoretical assumption of unlimited supplies of labor was crucial (1954, 1955). According to Lewis, the Caribbean has little to offer other than labor, and consequently, economic development of the region must be thought of in terms of how value can be extracted from Caribbean labor rather than on the demand for goods specific to the Caribbean (1954). Lewis argued that, because of a situation of surplus labor in the region, labor supplies far exceed the demand for labor. As a result, it is not demand that determines the cost of labor, but the basic necessities involved in sustaining a labor force. Under such conditions,

Lewis argued, it is possible to extract value from the labor force that in other parts of the world would need to be remunerated to the labor force (1954, 1955).

In effect, Lewis argues against neoclassical economic models that emphasize the relationship of supply and demand in determining labor costs, and adopts a classical economic theory of the extraction of value from labor. In so doing, Lewis uses a tradition of classical economic reasoning that runs from Adam Smith through Marx. Smith (1994) viewed labor as adding value to a product, with surplus value being found in additional labor beyond the work paid for by wages. For Smith, "Equal quantities of labour, at all times and places, must have the same value for the labourer" (1994, 36). Consequently, Smith's view of labor in relation to value was strictly in terms of its quantity and uniformity. As a result of Smith's emphasis on individual self-interest, laborers sold their labor as individuals, but paradoxically, diverse individuals had to be represented in uniform ways for equal quantities of labor to have equal values.

Marx criticized Smith and argued that workers were paid what their labor power was worth and at the same time added value to what they produced that could be extracted as surplus value. The key for Marx was distinguishing between the use value of labor as applied to the transformation of raw materials into finished products and abstract labor viewed as a commodity. The use value of labor refers to the value that the labor actually adds to an item produced. Abstract labor is labor as an exchange value—labor represented in an averaged and homogenized fashion so that it can be bought and sold. In abstract labor, the qualities of individual workers disappear, and the use of uniform time units plays a key role in the transformation of the use value of an individual's work into the representation of commodified, abstract labor. The worker only has labor to sell, and therefore must sell labor to meet his or her needs. Capitalists add value through the production process "[n]ot because his labour has a particular useful content, but because it lasts for a definite length of time" (1977, 308). Just as money becomes the commodity that can serve as a medium of exchange for all commodities, duration becomes a medium for measuring all labor and making it exchangeable. In Marx's economics, the extraction of surplus value is no longer something to be negotiated between an employer and employee, but between the class of capitalists and the class of laborers through the mediums of time and money.

Marx's model for the extraction of surplus value, and consequently Lewis' model of economic development, then depends on the price of labor. As Marx defines it, "The average price of labour is the average daily value of labour-power divided by the average number of hours in the working day" (1977, 684). In other words, this is an average divided by an average. Marx continues to say that the standard unit of measure for the time-wages from which the price of labor is calculated is the working hour (1977, 685). The surplus value consists of the hours worked over and above the hours that the worker works to produce value equivalent to the maintenance of the worker—this is the linchpin of Lewis' model of unlimited supplies of labor being able to entice industries to set up manufacturing endeavors in developing countries at no more than the cost of the labor force's basic needs.

In contrast to Lewis' model, neoclassical economic theory has criticized and discarded the surplus value of labor. Yet, the representation of labor in terms of duration remains a component of neoclassical economics—as I suggested at the beginning of this chapter, it is a cultural logic that transcends ideological differences. In a set of important publications beginning in the late 1950s, the foundation for neoclassical economic theories of labor was set by the study of the decline of work hours in relation to wages and consumption (Lewis 1957; Becker 1965; Boyer and Smith 2001). For instance, the work of Gary Becker has been described as "fundamental to the development of modern labor economics" (Boyer and Smith 2001, 208). In his important article, "A Theory of the Allocation of Time," Becker (1965) considers the "cost of time" in the relationship of working and nonworking time. For purposes of his analysis, he does not address different levels of productivity at different work times, or different levels of consumption at different nonwork times. Instead of relying on averages, as Marx did, he uses notation from calculus, such as T_w, to represent "time spent at work." While this approach allows for theorizing variation in the amount of time spent at work, it still defines it in terms of duration. Throughout his work, the cost of time is always the cost of duration.

SUBJECTIVITIES OF LABOR IN THE CARIBBEAN

Contrary to the ways of representing labor in economic theories, from a West Indian worker's perspective, labor cannot be reduced to commodified, homogeneous duration. The Caribbean is kairotic—or, as Benítez-Rojo (1996) would argue, rhythmic. The uniform progression of homogeneous duration is complemented by other cycles and rhythms. In Trinidad, no two years have the same rhythm of national holidays, since so many national holidays are religious holidays tied to religious calendars with different logics (see Birth 2013). In addition, there is a celebration of spontaneity in relationship to organization. This ethos is captured by writers such as C.L.R. James, Walter Rodney, and Franz Fanon. Rodney ends his book *A History of the Guyanese Working People* by contrasting the efforts of the colonial elite to organize labor in the face of workers' spontaneity and their ability to apply "themselves creatively within the limits of the local and international situation" (1981, 221). Fanon lumps the labor unions, the political parties, and the government together as entities that cannot grapple with the spontaneity of country people (1963, 123), and he attributes this spontaneity to the country people existing outside of the dominant ideas of order, organization, and scheduling that are part and parcel of colonialism. In *Notes on Dialectics*, C.L.R. James makes two important observations on this issue: first, that one must always relate organization to spontaneity (1980, 89), and second, that "the old categories have us by the throat" (1980, 17). These three West Indian thinkers point to a theme of the importance of significant, spontaneous actions, and, consequently, moments. They are not theorists of homogenous duration, but of spontaneous revolution.

In the Caribbean there emerges a contrast between the issues of how much time is devoted to an activity versus when the activity takes place—in other words, "how much" versus "when"—and this is related to representations of workers' agency and autonomy. The tension between duration and timing is part of the shared historic and cultural processes, or *oikoumenê*, that Mintz describes for the Caribbean (1993), and it can be heard in how Trinidadian workers talk about their work.

The Timing of Work

In rural Trinidad, it is often difficult to know the hours of a shop. Some rum shops are bound by the conditions of their liquor license to post their hours as "open any day, any time." But even small rural groceries do not keep regular hours. In part, this is because many shop owners also have other professions and tasks; in part, it is because some customers make demands at all hours of the day and night. As one shopkeeper told me, "I open when there are customers. When there are none, it make no sense to open."

This was driven home to me one night many months later. After midnight, I heard a man calling out in a loud voice to one of the owners of a nearby shop to sell him a cigarette. The shop was built underneath the owners' house, making them vulnerable to such late-night demands. "Betsy ... Betsy!" [a pseudonym] the man yelled, choosing to seek help from the wife rather than the husband. This would-be customer was one of the village madmen—a man publicly recognized as being mentally ill, but who was a functioning member of the community. Betsy and her husband had taken him under their care and ensured that he received his medication regularly. This night, he wanted a cigarette. Cigarettes can be purchased one at a time, and this man was outside Betsy's shop trying to wake her up in order to buy one cigarette.

After many bellows, I heard Betsy respond, "What you mean waking us up in the middle of the night?"

"I wan' a cigarette nah."

"The shop's closed!"

"Just open so I can buy one cigarette," he pleaded.

"Wait till morning," Betsy sternly replied.

"Come nah, I want a cigarette!"

"Okay, this once!" Betsy finally said with a sigh loud enough to be heard by the neighbors, and then she did the only reasonable thing—she woke up her husband to have him sell the cigarette.

The diversity of shop hours in Trinidad reflects the multiple uses of clock time. The clock can be used to determine when a shop opens and closes; or it can be used to determine how long a shop has been open; or it can be used to know the time, such as the ungodly hour of the night that some people want to buy cigarettes. In some contexts, the clock determines schedules, and consequently when action will take place, such as doctor's appointments or television programming schedules. In other contexts, clock time is a reference for parts of the day, rather than something to determine when people do a particular activity. A third use is to calibrate other definitions of time, for example, as one man pointed out to me when a flock of parrots was noisily flying overhead one evening, "It must be six o'clock—the parrots are flying overhead."

As a jack-of-all-trades, Bernard [a pseudonym], one of my neighbors in Trinidad, provided a sophisticated commentary on this contrast. He is a cocoa farmer and a carpenter, and he had a government job maintaining roads. One sunny afternoon, he was discussing with me the relationship of time and work when he worked for the government. The two ways in which such work is structured are day work and task work. Day work involves beginning to work at seven in the morning and ending at four, and task work, theoretically, involves enough tasks to equal about the same amount of time as day work. The conceptual structure behind both day wages and task wages is the same—it involves a concept of average labor power as measured by time; consequently, the wages Bernard received were the same whether he was paid by the day or by the task. This fits classical economic theory, but Bernard's thinking about work and time is not constrained by classical economic theory. His opening insight to me was, "Hours don't make work":

> Now I have always use this expression: "Hours don't make work." And I told a big boss that, my superior, an officer, a day. I say, "Hours does not make work." He talk about, you know, what working hours, a seven o'clock.

He wanted to make a point that the management of roadworkers undermined their productivity. In his next statement, he made clear that this was not a matter of laziness, but of using the knowledge that a "working man" has:

> I say, "I working man." He [the boss] talking about cutlassing to start at seven and thing about task and whatnot and sometimes talk about condition. I say, "A working man never wait on time." A real working man who like to do his work and thing, a man who has the spirit to work, the will to work, he never really look at time, right. When we reach to work, we working, let us say, a day work sometime.

With his passion for work established, Bernard argues that a "working man" does not wait until the boss says to start work, but begins to work upon arrival. In the tropics, one of the most common forms of roadwork is keeping the grass and undergrowth clear from the sides of the road, and anyone who has ever tried to cut such plant growth knows that it is much softer in the early morning than at midday. Bernard recognizes this, which is why he suggests that a working man begins work early. He then considers the consequences of this practice on the contractual workday:

> We don't really make seven to four. That is seldom that it happen. Most of them is maintenance workers and we on development and all kind of things. Even on development, we used to work seven to four, and yet as soon as we reach on work, we start to work. We never wait on time, to say wait till seven to commence work. Right? Sometime we reach before, we know that work we have to do, and before the overseer might reach or the supervisor

as the case may be, because we already started working already—we know what to do ... As soon as I reach, I start. Right? Although you might work to ten, half past ten to eleven. *From the time you start to work earlier, you will perform much more work than if you wait for the sun to come.* So you really doesn't wait on that ... The reason for that is that when you have this zeal in you to work, when you reach, you begin to work, so the amount of work you could do when the weather is cool, when the weather begin to get hot, you can't do it.

Bernard emphasizes that it makes more sense to begin to work when one can get the most work done. If one is "cutlassing," that is, cutting grass with a machete, then it is far easier to do in the early morning hours before the hot sun has dried and hardened the stalks and blades of grass. Along these lines, Bernard continued:

If we work from seven to nine and the sun is hot, we finish a task in two hours and the sun is hot. If you start at five when the sun is not hot, it might take less, an hour and a half, so by half past six you might finish. If you start at six, you might even finish by half past seven. So he [the boss] say a man have to start work at seven. I say, "But it is task we are doing, and that is the point I pushing here. We doing task—that is not day's work." If you work for the whole day, the foreman need to work seven to eleven, twelve to four on the sheet, but in task, the only time is seven and he inspect your task to see it's complete. If it's half past seven and you finish your task and you do the quantum of work he has no quarrel with you because you do the amount of work.

For Bernard it is not the quantity of time one works, but when one works that is important. Even though he refers to clock time to discuss his work, for him the clock is more important for determining the time to begin work than for the measurement of the duration of work. The task of cutlassing several hundred yards along a road when started at five in the morning is a smaller task than the same yardage of cutlassing when started at seven in the morning. His perspective of work relies on the use of time reckoning, in his case the position of the sun and secondarily the use of a watch, to determine when he should work, but his employer only uses time reckoning to measure the duration of work. In this way, the use of a watch is not only to determine when work should begin, but also to anticipate when work should end.

The conflict that Bernard faces with his employer is summed up by a recently retired worker: "You can get there at five o'clock in the morning, but anything that happen before seven, you get nothing."

Work Is for Working at Whatever Work Must Be Done

Bernard is aware of the use of the clock to measure work. Like many Trinidadians who discussed time with me, he suggested a moral dimension to the link between time and work. When one works, one should work well, and to work well often means working at the right time to get the work done. This is contrary to a representation of labor in terms of averages. For Trinidadians not all times and not all work are equivalent.

Rupert [a pseudonym], who used to work in an automobile factory, agrees. In reflecting upon his time in the factory, Rupert told me:

> You going work for an honest day for an honest day's pay and you know you come to work for that day you come to work eight hours. I believe that if you need work you supposed to ask for work. I used to do spot-welding work. I used to do a lot of maintenance work—anything you know. Like if I go, and it have no spot-welding that day and they say, "Go and push carts," you know, you meet a trolley or whatever.

Rupert was paid to weld—a skilled position—and when there was no welding, he chose to push carts—a less-skilled position than welding. One hour of a skilled laborer doing simple labor is one hour of simple labor, not skilled labor, but the extraction of surplus value from Rupert is based on the assumption that he is doing skilled labor. Classical economic theory of average labor-time stumbles in the face of Trinidadian labor practices.

Not only do Bernard and Rupert disrupt the implementation of one system of temporal orientation, but they also reveal a weakness in the concept of average labor-time. According to Marx and scientific management theorists, average labor-time represents the productivity of the average worker. What Marx does not acknowledge in his discussion of this average worker is that the productivity of the average worker is an average of the productivity of that worker over time. In effect, the use of labor-time as a basis for economic theory is the use of an average of the average work of average individuals. In scientific management practice, supervisors would choose one worker to represent the average. Even when turning to companies that sought a more humane approach to scientific management than that of Taylor, one finds this problem. For instance, confectionary company Rowntree's was known for its social welfare approach to its workers, in part because of its Quaker roots, but also because of the interest B. Seebohm Rowntree had in the well-being of workers—an interest that led him to be a leading social reformer and sociologist studying social welfare in addition to being a company executive. The labor manager of Rowntree's, C.H. Northcott, wrote of:

> [the] company's general policy and practice to take as the subject for time study, works of average ability and skill, in whose selection the shop stewards were jointly responsible with the management. Wage records were helpful in establishing the average standing of workers, but an operative with a wage equaling the average of the section did not automatically become a subject for selection for time study. Other factors such as temperament and consistency of performance were taken into account. (Northcott 1956, 339)

Northcott then adds, "Should any dissatisfaction arise subsequently, the management was required to prove that the rate of working during the tests could be maintained under normal conditions" (1956, 340). Parsing Northcott's approach, the representation of the average worker used to set piece rates was not necessarily the worker who worked closest to the average rate of all workers, but a worker of the right "temperament" and "consistency of performance" on whom management and the workers could agree to be studied as an average worker.

My Trinidadian friends, Bernard and Rupert (whose ancestors supplied chocolate to Rowntree's), do not see work through the lens of finding an average worker to represent all workers. To them, workers are unique, and the timing and rhythms of their work play a role in defining their uniqueness. In interviews, I found indications that this resistance to averages emerges as early as when children are in school. "Cindy," a young woman whose siblings still attended school, and who had only recently finished school when I interviewed her, told me that eagerness and discipline in school varied "according to the student," however her contrast was not built on conduct during regular school hours, but on the basis of "who remain after hours to go to school." For her, the eagerness to learn was indicated by the desire to ignore the school schedule and the timing of educational work.

In fact, there are several idioms that Trinidadians use to obscure duration: "jus' now," "Any time is Trinidad time," and "long time." In *Any Time Is Trinidad Time* (Birth 1999), I described how these can be used as glosses to hide actual durations. In the workplace, the goal is to be able to convince the boss that the workers have worked the duration management desires when the workers have not. In the stories I was told, this was not for lack of hard work—quite the contrary, the typical tale was of a worker who had finished his task before his boss thought he would, and then would seek to represent the duration he had worked as what the boss would expect. In Trinidad, the connection between duration and work seemingly has been negotiable for a long time in practice, although such negotiations are not permitted in policy.

Differences Between Workers

It is clear that neither Rupert nor Bernard thinks in terms of socially average labor-time; in fact, Bernard, in particular, seems to think against that concept. Not content with his critique of average labor-time through his discussion of timing, Bernard also points out differences between workers employed for the same job:

> Now a man will stay up county work at seven o'clock, right, and he will break off at four, right, working. And you see constant on the job. Right? Same condition, another man will come to work, meet that same man there working, he will come one hour after and finish work at three, will get more work from that same man. So you see, it is not the hours. It is the amount the man can put down under the same condition.

The concept of average labor hides such differences, but from the perspective of classical economics, this does not matter. In fact, the point of representing labor in terms of averages is to be able to assign a value to it and thereby allow it to be represented in terms of its exchange value rather than its use value. Consequently, the management of underperforming workers is not solely to increase their individual productivity, but to improve the overall average.

The concept of average hides a diversity of productivity across workers and the diversity of productivity at different moments in time. The nature of averages is that they are most meaningful when calculated from a large number of cases in which there are no outliers. As representations, averages homogenize diversity. Bernard resists being represented in such a way. He asserts his personhood as a worker, and in so doing, he challenges averaging his qualities in a mass of homogenized labor. The importance of work to Bernard's sense of who he is relative to others is not unusual. As I shall argue in the next chapter, it is concepts of labor and personal labor histories that structure how Trinidadians represent their past. In effect, labor is crucial to structuring memories and presenting oneself to others. The homogenizing quality of the representation of labor in classical economics conflicts with Trinidadians' efforts to describe their individuality. As Marx writes of labor-time, "It is the labour-time of an individual, *his* labour-time, but only as labour-time common to all; consequently it is quite immaterial *whose* individual labour-time this is" (1970, 32). The previously discussed method of Rowntree's labor manager, C.H. Northcott, in choosing an average worker for time studies exposes the flaw in Marx's model—it can matter whose labor-time is used to define the average, and how the average then becomes defined is an act of managerial power, not merely an observation of fact.

Bernard argued that workers are not equivalent and the effect of timing on many jobs makes durations not equivalent. In contrast, Marx's theory uses averages in a way that hides the diversity of workers and the importance of timing. Likewise, neoclassical economists also hide the diversity of workers and timing by opting for mathematical differentials that assume that the variations within the defined limits are differences of quantity and not quality. By saying that workers are very different, Bernard challenges the position which holds that Marx really meant an average of the same work over multiple workers, and he challenges the idea that workers can be represented solely in terms of the quantity of work they perform. By saying that the effect of the timing of work on the work itself affects productivity, Bernard challenges the validity of an average of the same workers' work over a period of time and completely undoes any ability to represent work in terms of duration. Whichever averaging algorithm one chooses in economic models' equivocation between the average work of one individual or the average work of a group of workers, or whatever differential limits economic theories employ, Bernard has a temporal orientation system that undoes them, and in so doing, he challenges the homogenization of labor through consciousness of labor.

A further examination of what Bernard says fosters another interesting discovery: He postulates one man working eight hours (nine hours minus one hour for mandated breaks), and one man working six hours (seven hours minus one hour for mandated breaks). The use value of the labor of these two workers is equivalent. The average of these workers would be seven hours per unit of work, but the represented average, according to the County Council, would be eight hours per unit of work. In other words, in Bernard's example, the average calculated from the workers' labor is different from the average assumed of the workers by the government agency that employs them. This disjunction is often pointed out by those who comment on government work in Trinidad. For instance, Krishna [a pseudonym], a prominent Indian man in the community, complained of how the advent of the unions combined with political parties' desires to court union votes has resulted in government union workers getting paid for more than they work—they are paid for eight hours, but work fewer than eight hours. In roadwork, anybody who takes more time to complete their work than the time allowed by the supervisor tends not to be hired into a permanent position. So, in fact, the eight hours cannot be a statistical average of workers, but instead functions like the upper limit of a range of work.

From the point of view of the representation of the value of labor, this either indicates a situation in which the pay, even though it is attributed to eight hours, really represents the value of labor that occurs in seven hours, thus making the eight hours a fiction, or that the wages for eight hours and the agreement that workers would work for eight hours represents a former relationship of either the extraction of surplus value relative to the cost of labor (classical economic theory) or a former balance of labor supply and demand (neoclassical economics). In either system of economic theory, laborers working for seven hours while being paid for eight is a sign of inefficiency in need of a correction.

The outsider, predominately neoclassical perspective of the World Bank and the International Monetary Fund (IMF) in fact implicitly adopts the latter interpretation, and uses it to justify cuts in the public service workforce as part of economic structural adjustment. Yet, it is clear that the ideas of labor shared by Bernard and Rupert do not follow the logic of standard economic models. For these men, labor has to do with issues of timing and personal qualities that defy the ability to average labor over homogenous time: time is not homogenous, and workers are not equivalent.

Thus, Trinidadian work practices challenge classical economic assumptions in three ways: first, by holding different jobs at which one receives a different rate of pay; second, by emphasizing the timing of work rather than the duration of work; and third, by recognizing that some workers are better than others. It is not an emphasis on absolute individuality, however. Bernard compares workers. Workers' qualities exist in relationship to other workers' qualities, and the qualities of particular times of day exist in relationship to the qualities of other times of day.

Multiple Rhythms and Kairos

Combined, this polyrhythmic challenge to the homogenous durations on which classical and neoclassical economics are premised is an assertion of relational qualities and kairotic time—time represented in terms of ideal moments rather than exchangeable durations. If government work is viewed according to these two dimensions, then it is clear that the point of government work is not concerned with representations of average labor in order to commoditize it or with an equilibrium of labor supply and government demand for labor that determines wages. In fact, government employment practices emphasize patronage—developing personal loyalties. This was actually a crucial feature of the development structure initially implemented by the government in Trinidad (Ryan 1972; Hintzen 1989). This structure involved a local-level village council which was to manage local projects that would be funded by the government. In the original conception, as articulated by W. Arthur Lewis, the government would provide materials for infrastructural improvement and the village council would organize a volunteer labor force motivated by the desire to improve their community for their own benefit and to compete in the Better Village competition (Craig 1985). Whereas this was the model, I find no indication of the voluntary labor program ever occurring in the community in which I conducted research. Instead, the village council came to be looked upon as a source of employment opportunities associated with local patronage. As one former councilmember told me, he would learn of local projects from the County Council office and would be asked to gather up a workforce of temporary laborers. Typically, these were five- to 10-day projects, but provided a much-needed source of cash for local households.

This form of employment is open to charges of abuse. In 1996, just before parliamentary elections, the rumor in the village was that the ruling party had told the local village council to only hire loyal party members. Other times when the availability of temporary labor is high are before Christmas and before Carnival. The logic that drives government employment patterns is, thus, *timely* patronage, and not patronage without reference to timing—in effect, kairos. As argued at the beginning of this chapter, kairos contrasts with an emphasis on uniform duration. Moreover, the emphasis on proper timing inherent in kairotic approaches has political dimensions.

Timing influences the availability of government work, with more employment being offered before elections, before Christmas, and before Carnival than at other times. It is well known that in Trinidad the number of government projects that employ temporary laborers increases as elections approach—this is a kairotic cycle of employment that suggests that the use value of employment must include political benefits as well as any infrastructural improvements that result from the labor. Furthermore, employment is viewed as a form of patronage—an exchange of a job for a vote. Eventually, if a worker accumulates enough temporary days, the worker is made permanent. Consequently, combined with any logic from economic theories for understanding labor is another dimension that emphasizes timing and the cultivation of political sympathies. In fact, this is a long-standing feature of Caribbean labor that intertwines patronage and timing with work.

Consequently, in Trinidad, labor is not simply conceptualized as an exchange value, but as a set of political and social relationships that are never fully hidden by the commodification of labor. These extra-economic characteristics create a subjectivity of work that refuses to be commodified, and incorporates dimensions of personal liberty and politics in ways that are distinctively West Indian. These dimensions become highlighted as one shifts from a logic of representing work in terms of homogenous time toward representing work in terms of the timing of tasks and relationships—a shift from thinking in terms of measured duration to thinking in terms of kairos.

Marramao is careful to point out that chronos refers to numbered time but not to measured time (2007, 8). This insight reveals a common conflation between number and measure that the current cultural logics of time make. An example can be found with the concept of the year and the mundane task of reporting one's age. Age is *numbered* in years, but it is assumed that the number of years indicates the measure of one's age. In fact it does not, because years are not of equal length. Some years are leap years, and some years are not. Even so, the cultural tendency to conflate the numbering of time periods with the measurement of time results in years being treated as units of measure.

Kairos embodies an idea of time that emphasizes "personal factors in the situation" (Robinson 1968, 58)—this seems much more similar to the sense of time represented by Bernard than economic theory's concept of labor. Tillich is not content to merely contrast kairos and chronos, but uses it as the basis for a criticism of occidental concepts of time. He claims that Western concepts of time are built on the model of the machine; they have created "a rational conception of reality as a machine with eternally constant laws of movement manifest in an infinitely recurring and predicable natural process" (1948, 34). Tillich prefers the concept of kairos, a concept of history that embraces every moment as a moment loaded with meaning rather than placing moments in the march of time (chronos). This is affirmed in Sian Lazar's discussion of the temporalities of street vendors in Bolivia—for these vendors there is a contrast between a repetitive temporality of daily protest and a temporality of historical narrative. The interaction of these temporalities creates the possibility of "discontinuity or rupture" (2014, 106). Tillich resists "any attempt to absolutize *one* historical phenomenon over against all the others," and he resists "the leveling of all epochs into a process of endless repetition of relativities" (1948, 42). He wishes to combine the uniqueness of each moment with the general idea that every moment is important.

Based on the contrast between the ethnographic evidence about work and economic models of labor, one might conclude that the importance of timing has a certain precapitalist, agrarian sense about it. In agriculture, so much of what is done must be done at the right time. Critical in Marx's discussion of the commodification of labor in industrial capitalism is the use of time rather than skill or productivity as the means to represent labor in terms of value. Variations in productivity, say as fatigue sets in, are irrelevant to capitalist calculations of labor value. Yet, despite E.P. Thompson's thesis that workers adopted clock time to represent their labor (1967), timing never went away. Despite the use of the clock with its homogenized hours to organize time, and to suppress the differential value of different moments, timing has remained an intrinsic part of the human experience of time. There is still a time to plant and a time to reap. Even in manufacturing there is seasonality—for example, the production goals to get product to store shelves for the Christmas season. In the belly of the capitalist beast currently sits the importance of high-frequency trading—the use of computer-driven algorithmic trading that occurs with timestamps with a precision of nanoseconds and trading cycles that can occur thousands of times per second. In high-frequency trading, timing is everything. Moreover, distribution systems are seeking to cut delivery times, and hence, the duration of the circulation of production, commodities, and capital that Marx discusses in volume 2 of *Capital* (1978). Paul Virilio describes this phenomenon as the emergence of "just-in-time distribution systems" (2010, 33). A difference of a millisecond or less in executing a trade makes the difference between profit and loss. Securities trading is kairotic.

All of these different perspectives provide corroborating cases on the unnaturalness and antihuman character of the concept of average labor-time. The challenge to the representation of "average labor" and the related idea of "socially average labor-time" (in classical economics) has subjective significance beyond the ways in which it undoes economic theory. The target of Bernard's critique is not labor economists but his boss, and his audience was the American anthropologist interviewing him—a contrast based on power and position and a contrast based on conceptions of national culture. The two become intertwined in important ways.

Time and Agency

In many interviews I conducted about conceptions of time, Trinidadians drew contrasts between themselves and the United States—even individuals who had never been to the United States. The contrasts were consistently over the relationship of time and work. Erwin [a pseudonym], a young unemployed former factory worker, told me:

> The majority of Trinidadians almost think alike. They ain't afraid. They want more fete than anything else. They enjoy themselves freedom ... The system what have overseer and them does start work with it because in another country it entirely different. Like if you go America, you jus' working, it have more work than play. Down here, you playing right through—you working and you playing. You go to work and leave your work to go a fete, you go out and that sort of thing. America like work is important. Down here, the majority of people finish work down here eight o'clock. They go to work seven o'clock and they finish eight o'clock, and they getting pay for eight hours. America is not like that. You get paid for an hour, you understand? That is not so down here. Down here the system real different entirely to the whole part of the world.

At the same time, representations of privilege in relationship to time emphasized individuality—the rhetoric employed by Bernard defies homogenizing time. To have individualized time was to have power, prestige, freedom, and independence. This is not something new. Based on his study of this same village in Trinidad in the late 1950s, Morris Freilich notes that occupations are ranked by the village residents in terms of the independence they offer the worker, not in terms of pay (1960). In the ranking he generated, independent farmers were ranked higher than schoolteachers. The linking of temporal sensibility and independence is something I did not completely recognize when I wrote *Any Time Is Trinidad Time*. As one man put it, "Trinidad time is any time, oh hoh, what they mean is the freedom: they leave anytime, they go anywhere, you know."

In an important way, then, labor becomes tied to conceptions of independence, and independence to a sense of individuality. Consequently, the project of reclaiming one's labor is part of a larger project of individual expression. It marks one dimension of contrast between the United States and Trinidad, but a subjectively significant one both in terms of its representation of Trinidadians being subordinated by outside economic ideas and in terms of representing Trinidadian character.

The contrast of chronos and kairos, then, represents the contrast of homogenous impersonal time versus agentive personal timing, and as a consequence, the contrast between the Trinidadian experience of work as a reflection of individual agency and the global hegemony of average labor. It is a contrast between the abstracting quality of measuring activity in terms of duration, and the personalized quality of measuring time in terms of personal action. If, like Postone (2003), one views Marx's theory not as a theory of labor, but as a critique of labor, and then views the emphasis on average labor-time as part of what is meant to be critiqued—as one of the sources of alienation—then the contrast between chronos and kairos suggests that part of what is alienated is the personal sense of timing. This is surely what Bernard suggests in his resistance to how his boss represents work.

CARIBBEAN TIME AND CAPITALISM

Mintz suggested that industrial capitalist work organization originated in the Caribbean (1985, 51)—that the factory system so crucial to the temporal reforms that E.P. Thompson identifies (1967), so foundational to the classical economic theory of the extraction of value from labor (Marx 1977), and so foundational to the representation of the supply of labor in neoclassical economics (Becker 1965) was forged on the sugar plantations of the West Indies. Yet, it seems that as abstract, average labor was created for the plantations—with the enslaved Africans becoming the embodiment of commodified labor—the counterpoint of an alternative way of experiencing labor also emerged. In a strange parallel, the models of the value of labor found in post–World War II development economics were also forged by the practical knowledge W. Arthur Lewis brought from the West Indians combined with his grasp of the economic theories and debates of his time (Tignor 2004). Outside of the abstracted averaged labor-time was the kairotic relational labor-time. Timing and relationships, meaning and intersubjectivity, patronage and competition coexisted and challenged the capitalist culturally shaped economic logic. With independence, the emphasis on timing and patronage became embedded in government employment practices, and in a nation such as Trinidad and Tobago, where the government used petroleum revenues to seek 100 percent employment, there has emerged a labor force that is unruly in relationship to capitalism because it is formed on a conflicting logic of kairos and sociality.

Viewing time as consisting of units to be filled or as units by which all activities and processes can be measured and thereby represented is one temporality among many, although a temporality frequently privileged in economic analysis. For instance, as Bear (2014b) points out, Piketty's recent discussion of capitalism (2014) emphasizes this temporality of time as a container because it is amenable to quantification. Bear (2014b) rightly argues that Piketty's assumptions about temporality elide rhythms and temporalities that can produce uncertainty and insecurity. She calls for a qualitative approach to the study of capitalism, in part so these temporal conflicts can be revealed.

Homochronism seems to be a necessary condition for the commodification of labor. This obscures the experience of work in which proper timing greatly affects the ability to quickly complete a task. What homochronism allows is a fungibility of labor from one person to another or one context to another; what it erases are all the sources of variation. Just as with the representation of the past in which not all years are equivalent, which will be discussed in the next chapter, so seems to be the case with work in which not all moments are equivalent. Homochronism obscures the importance of timing in labor. The economic perspective on labor, then, just as with ethnographic description, psychological studies of timing, and chronobiology, is distorted by the lens of uniform, homogeneous time.

In his book *Imagined Communities*, Benedict Anderson places homogenous empty time among the "cultural roots" of nationalism.

Anderson's idea of temporality provides a phenomenological foundation for his overall theoretical framework. The shift in temporality he describes was a shift in "modes of apprehending the world" (Anderson 2006, 22). Wickman suggests: "'Homogenous, empty time' signified a temporality purged of foreshadowings and fulfillments (such as those prominent in the Judeo-Christian tradition), with time measured instead by clock and calendar" (Wickman 2007, 100), and he adds that "empty time enables a hegemonic mastery over the future by the present, and as a consequence over social peripheries by a dominant core, by subjecting contingent events to calculations of probability that conform to a logic of investment and return, and hence that infinitely extend the socio-economic axes of power deployed by advanced nations (e.g., Britain)" (Wickman 2007, 101).

Bhabha and Kelly point out Anderson's misappropriation of Benjamin's ideas and that Anderson's idea of homogeneous empty time belies the hierarchical power relations on which it is built (Benjamin 1968; Bhabha 1994, 161; Kelly 1998). In another criticism of homogeneous empty time, Chakrabarty (1997) argues that this concept of time treats time as belonging to nature itself rather than as the hegemonic construction it really is. These criticisms point to a conceptual danger of treating the uniform temporality of homogeneous empty time as natural or even as constitutive of modernity. While Anderson gives credence to the arbitrariness of the sign, his discussion of homogeneous empty time slips between time markers as signifiers and an ontology of time that naturalizes and universalizes homogeneous time. This ontological problem of time is felt if one brings Einstein's theory of relativity into the mix—if, as Einstein argues, time is relative to motion and space, then a time that is uniform and homogeneous across space is a fantasy, and simultaneity across space is in Einstein's own view an "illusion" (Einstein 1992, 378). If this is the case, the signifiers of dates and clock times are not only arbitrary, but refer to a signified that is also arbitrary.

CHAPTER 4

Past Times: The Temporal Structuring of History and Memory

Over recent decades, the study of the telling of history has emphasized the relationship of narrative structure and history. Inspired by the work of Hayden White (1975), this literature demonstrates the diverse ways in which the past gets represented. In the Caribbean, Richard Price (1990, 1998) has explored cases of the same events producing very different historical narratives. David Scott has shown how the dominant narrative structures in the writing of Caribbean history have also shifted from an emphasis on romantic narratives to a recognition that tragic structures are more appropriate (2004).

While the work on the cultural and narrative shaping of history has burgeoned, it has frequently been disconnected from discussions of the temporal assumptions of modernity that shape historiography, despite explicit statements that relate time and narrative (Ricoeur 1984, 1985, 1988; Kermode 2000; Bakhtin 1981; West-Pavlov 2013). Instead of narrative temporalities, discussions of time in relationship to modernity emphasize "homogeneous empty time," or homochronicity. While this concept was developed by Benjamin in explicit contrast to a view of time with a narrative purpose (1968), its deployment in social theory has been to document how this temporality has been crucial to the construction of ideas of contemporaneity that allow for the imagination of communities across space (Anderson 2006). More recently, this homochronic view of time has been treated as playing an important constitutive role in the emergence and persistence of secularism as a cultural system (Taylor 2007; Asad 2003).

The application of homochronic time to memory and history changes them. As Eric Hobsbawm observes, "The moment when real time is introduced into such a past (for example, when Homer and the bible are analysed by the methods of modern historical scholarship) it turns into something else" (1997, 22). Braudel makes a similar point: "[S]ocial time does not flow at one even rate, but goes at a thousand different paces, swift or slow, which bear almost no relation to the day-to-day rhythm of a chronicle or of traditional history" (1980, 12). If, in contrast to the academic study of the course of events, we are seeking to understand how local memory and tellings of history shape people's ideas about the present and future, then this imposition of chronology and the traditional tropes of academic history obscures more than it enlightens.

Homochronicity allows the creation of linear chronology as a means of organizing history. Charles Stewart writes that this temporality of historical consciousness "is a culturally specific Western development that arose in a particular period in a certain area of the world" (2012, 2). Seeing time as this linear string of homogeneous units stretching from the past into the future that are filled with events organizes not only how history is commonly presented, but also how the future is planned. Alternative temporalities, such as those generated by the interaction of dreams, memory, and history that Stewart (2012) documents in Naxos, Greece, do not fit into homochronic chronology. As he argues, casting them merely as resistance to dominant ideology impoverishes their role in the imagination and agency that are particularly important in thinking about crises.

There is an uneasy coexistence between homochronic chronology and the weaving together of events in myth and narratives to tell stories about the interaction of the past, present, and future. Homogeneous empty time does not support narrative structures. Narrative involves the interaction of beginnings, middles, and endings (Kermode 2000), whereas empty homogeneous time extends infinitely into the past and the future—it has no beginning or end. Narratives emphasize some moments as more important than others, whereas homochronicity treats all times as equal. Moreover, as West-Pavlov argues, narrative constitutes temporalities (2013, 83ff.).

Anderson attributes the ascendancy of homogeneous empty time to the rise of nations in the eighteenth and nineteenth centuries, but this period also involved the emergence of tensions between narrative temporalities and homochronicity. About this period, Alkon notes that the consciousness of the authorial construction of time in narrative seemingly bursts into consciousness, with experimentation and innovation proliferating afterward (1979, 16). The explicit focus on time in the works of Baudelaire, Poe, and Verne discussed in the Prelude is an indicator of this trend.

This chapter explores how time is represented—indeed, warped from a chronological perspective—in narratives of the past. It adopts the perspective of what Zerubavel (2003) calls *time maps*—the manner in which memories become temporally structured. Points in time that organize these maps are temporal landmarks. As with the contrast between kairos and uniform timekeeping, this is a contrast between the meaning of moments and homogeneous chronology. What will be described in Trinidad are patterns that structure the past and make tales about the past in the present intelligible. Yet, the intelligibility defies simple chronology. Moreover, the temporal landmarks pull events toward them, so that events cluster rather than getting a chronologically correct distribution across the years. As Joanne Rappaport states, the representation of history is a "choice of a particular expository style that is itself determined historically" (1998, 13). Following Bakhtin (1981), an implication of Rappaport's insight is that representations of history involve choices of chronotopes, and these chronotopes may or may not match the chronotopes typically used in scholarly historical representations.

Time maps are as applicable to personal narratives as they are to historical narratives. The study of episodic and autobiographical memory shows that subjects' use of temporal landmarks to structure their memories has become an important issue. But in psychology, the explanation for why some events are chosen as temporal landmarks and others are not has emphasized individual cognitive efficiency and effectiveness (Barsalou 1988; Brown and Chater 2001; Brown et al. 1986; Conway 1992; Robinson 1986). These explanations include ideas that regular, predictable events are more useful than events that are not (Shum 1998); that traumatic memories serve better as landmarks because they are crucial reference points for one's autobiography (Berntsen 2001); and that humans develop internal calendars that serve as temporal landmarks (Kurbat et al. 1998).

These are not satisfying explanations for many reasons. First, not all temporal landmarks are regular, predictable events—the attack on Pearl Harbor (White 1999), being a prisoner of war in Asia during World War II (Murakami and Middleton 2006), the bombing of Dresden (James 2006, 2012), and the 1947 Malagasy rebellion against the French (Cole 2001,

2006) certainly do not fall into the category of predictable, regular events. The Oil Boom in Trinidad and the period of prosperity of the Santa Fe Mining Cooperative (Ferry 2005, 2006) were not traumatic events, at least for many. An event as complex as the Chicano movement of the 1960s and 1970s does not easily fall into any of the proposed functional, psychological explanations—its significance seems profoundly interpersonal rather than resting on just individual memory function (Vélez-Ibáñez 1996). Finally, none of these events indicates an internal calendar.

Explaining temporal landmarks based on their function does not explain why memories cluster around such events, or why the landmarks themselves often cluster. Instead, the logic that psychologists use parallels that of classical economics. It postulates uniform time, and if time is uniform, then regularity is the most efficient principle for organizing memories of events. The temporal organization of memories defies the assumptions made by psychologists.

In contrast to emphasizing internal, cognitive function, I shall argue that it is the processes of making sense of others' presentations of their memories and of making one's own memories intelligible to others that drive the use of these landmarks. The use of landmarks is part of the dialogic, socially oriented process of crafting one's identity. If landmarks were solely for individual cognitive function, then there would be no need for the landmarks to invoke powerful intersubjectively significant means of establishing shared knowledge, and there would be no sign of cultural variability in terms of what makes a memory suitable to be a temporal landmark. On the other hand, if the landmarks are for the establishment of intersubjective significance, then they should be profoundly social and there should be evidence of cultural variability. This chapter studies the ways Trinidadians temporally structure their constructions of the past and discusses how these temporal structures strongly invoke significant moments in the labor history of the community and island. Instead of uniform chronology, one finds narratives with lacunae in events and periods of great importance that exert an organizing influence in the narratives.

To collect discourse on concepts of time, I employed nondirected interviews that focused on temporal idioms used by Trinidadians such as "Any time is Trinidad time," "jus' now," or "long time" (see Birth 2004). I used my community-wide census to generate a sample stratified by gender, ethnicity, and occupation, and eventually conducted a series of nondirected interviews on concepts of time and life history with twenty individuals, and single interviews on "long time," life history, or village history with an additional seven individuals. My observation of daily life in the community provided abundant cases of disputes over timing of activities that I used to prompt discussions in the interviews. By asking informants to give cases of the use of temporal idioms and to comment on disputes that arose when such expressions were used, I obtained a wealth of information not only on time, but on many dimensions of life. Of these temporal idioms, "long time" is the most important for exploring representations of the past. It has two related meanings: it refers to the distant past, and it is used as a modifier to emphasize the distance of past events. The query "Tell me about 'long time'" would often lead to rich, detailed recollections in which personal experience intersected with events of historical importance. For the most part, my interventions in these interviews were intended to get informants to expand upon what they said, rather than to fill historical gaps in their narratives.

The use of "long time" as a modifer emphasizes separation between a past event and the present, for example, "The road was paved long time," or "I finish work long time." Old heads, as elderly Trinidadians are called, are particularly interesting expositors of the multiple meanings of "long time." Mr. Roopnarine [a pseudonym] was one such old head who willingly and frequently allowed me to serve as his audience. By the time of my arrival in the rural Trinidadian community of Anamat in 1989, he was retired from the Ministry of Public Works, and was supported by his pension and his adult children. He daily relaxed on his front porch (or *gallery* as Trinidadians call it), or on a bench in his carport. As people passed, he greeted them, and if they had time to stop, he would engage them in conversation. One day I was interviewing him when we had the following exchange:

K: For you, how long ago is "long time"?
R: For me? Well, this is a funny question, you know? How long, well, for me I would say, if I were to say, I would never say "long time." I would say "long ago."
K: When your brother says "long time"?
R: When my brother says "long time" what he's referring to, well, I get the impression that might be 20 years or 30 years aback, or things like that. Or otherwise, I might even ask him what was a long time or something—what year that was? He might say, "when I was young," or "when I was ten." He say, "When I was 15 years I did this—that was a long time."

Several weeks later, when I was interviewing Mr. Roopnarine's brother, I asked, "What does 'long time' mean when you use it?" and he quickly answered, "Long time is when I was a younger fellow."

Young adults either corroborated old heads' definitions of "long time," or they emphatically deferred to the old heads' opinions on the matter of "long time." "Marvin," a man in his 30s, told me, "Most of the time, the 'long time' mean years away, that is years back." "Cynthia," a teenage girl, said, "It come like younger people wasn't around in that space of time, but to mean when they say 'long time,' they talk about the forties, the fifties going back. I wouldn't say the sixties because it have youths now born during the sixties, and most of them from the sixties come up."

In many interviews, another way of talking about what happened "long time" became apparent. Roger, a Spanish-Creole man in his 50s, was discussing with me how references to the past were often made for the sake of comparing the past to the present. The example he offered was about roads:

> People make a comparison with this road [the road running in front of his house], [and] many roads that were developed, to the Churchill Roosevelt Highway, which we call the MP Road. And they tell you since in the base days, which is 1940 or thereabout, that road was built and it never got no major repair.

The term "base days" refers to the massive American military development of Trinidad during World War II. In this statement, Roger offers two temporal frames: a year, "1940," and "the base days," a reference to a period of historical significance for the entire island.

Embedded within Roger's opinions are different ways of reckoning time in the past. "Long time" might mean "20 or 30 years aback," referring to a reckoning of time in years in relationship to the present. "Long time" might refer to a specific year or decade—an absolute chronological marker, for example, "1940," or "When they say 'long time' they talk about the forties or fifties." "Long time" might refer to when the speaker was of a particular age or life stage, for example, "when I was 15 years," or "when I was a younger fellow." Finally, it might also refer to a historical period, for example, "the base days." From these statements, one can infer four ways of structuring time to represent the past: (1) reference to the years separating the past event from the present, (2) reference to a date, (3) reference to a stage or age in one's life, and (4) reference to an event or period publicly recognized as historically significant. The relationships and use of these temporal landmarks was of greater concern than their accuracy, although documentary sources were used to determine the dates of historical events used as landmarks.

Most approaches to the study of memory and history focus on a significant period or historical event through bringing together documentary evidence and ethnographically situated recollections. The memories I am discussing here were gathered in a different fashion for a different purpose. The goal of my interviews was not to study history, but to explore cultural ideas of time. In these interviews, Trinidadians used memories to make points. Memories also served as responses to my prodding informants to expand upon statements they had made, or to offer examples. For instance, I was interviewing a middle-aged man about the uses of the phrase "Any time is Trinidad time," when the following exchange occurred:

K: Can you think of any specific times that you have heard it ["Any time is Trinidad time"] used?
M: No, not right now, but is long time, you see, is mostly when I heard this type of thing used was mostly when I used to associate more liming and thing—like the people on the block—associate myself more when I was working Amalgamated amongst plenty people; you know these thing. They would use this assertion.

So, in response to my asking for an example of the use of a phrase, this individual first referred to "liming"—a word that encompasses loitering, attending parties, and spending time with family and friends. He then associated liming with a specific location in the village called "the block," where young men congregated. Finally, this individual referred to a period during the 1970s and early 1980s when he worked in Amalgamated's automobile factory, at a time when he was a young man. In the interview, I was not seeking his memories about this period, but that is where he led me as a major source for his examples. Throughout the interview material, four cultural temporal models emerged: (1) dates, (2) periods of publicly recognized historical significance, (3) age or stage in the life cycle, and (4) reference to "years ago." Of these four, the first three are commonly combined.

The memories offered under these conditions seem patterned—they are not of events randomly distributed throughout the past, but manifest clustering. Such clustering of remembered events is not a particularly Trinidadian trait—Halbwachs noted it, as well:

> Why does society establish landmarks in time that are placed close together— and usually in a very irregular manner, since for certain periods they are almost entirely lacking—whereas around such salient events sometimes many other equally salient events seem to gather, just as street signs and other signposts multiply as a tourist attraction approaches? (1992, 175)

If the recollected events are charted by means of the cultural structures of time Trinidadians use to represent the past, not only is clustering apparent, but also so are some of the ways in which these ideas of time interact to relate individuals to collectives, and locales to the nation and the globe. It is not by accidents of history that remembered events cluster, but by the complicated work of relating self to others. I shall apply these ways of structuring time that emerged from the interviews to chart the memories that were also offered in these interviews. Even though most memories were only represented through one or two temporal structures, most can be charted in relationship to dates, stages in the life cycle, and events of historical significance. Admittedly, this is not their original form. The justification for this is that these three types of temporal organization do frequently co-occur, whereas the fourth type, "years aback," does not typically occur with the other three. The frequent co-occurrence suggests a possible cognitive relationship worth exploring. The subsequent discussion and graphs show that underlying these three of the four models of time are other principles of structuring narratives and relating autobiography to history, individuals to collectives, and specific locations to the country or the world. The fourth structure, "years aback," seems to create a specific type of narrative different from the others.

Years Ago

Mr. Roopnarine clearly indicated that using the phrase "years aback" is one possible way of temporally structuring representations of the past. This was a common response to queries about the meaning of "long time," but while it was used to define the term, it was rarely combined with the other ways of temporally structuring the past—dates, events of historical significance, or life stages.

When the temporal reference "years ago" was used, it was applied to morally tinged anecdotes—tales of the past that had to do with tragedy or scandal. For instance, one such anecdote told to me by an elderly Hindu man was:

> I know an instance is where your mother and father would not consent for the son and the girl to marry. That happen, even in St. Augustine. A few years ago, they both went in a certain lonely place, and both of them took poison. The boy and the girl both. When they find them, they find both dead.

The purpose of the tale was to explain to me how it was risky for Hindu parents to insist on arranged marriages for their children when those children were already in love with somebody.

Even stories that are datable and part of local history are temporally framed as "years aback" if there is a moral message. Stories about abusive overseers and local personalities adept at tricking employers and eluding the law took this form. Most striking in terms of its temporal framing was a story about the shameful and tragic demise of three brothers who supposedly entered into a deal with the Hindu goddess Kali for wealth. These brothers became wealthy, but then died painful, early deaths. The story as related by Hindu old heads carried a warning against dabbling in such religious practices. This story should have been easily related to years or life stage by one of the men who told it to me—not only did he inherit some of the land from these brothers, but with regard to all the other land in his possession, he always specified the year in which he acquired the land, thus making the "years aback" framing of the story about his relatives seem anomalous.

The use of "years aback" only approximately places stories in the past, and the fact that such stories are not explicitly linked to dates, stages in the life cycle, or events of historical significance gives these stories a timeless quality that rhetorically reinforces their didactic messages. "Years aback," then, is a temporality outside of chronology and is used to make points that transcend moments in time and even epochs.

Dates

Using "years aback" or "years ago" differs from using dates. These phrases explicitly emphasize a past that is separated from the present, whereas the use of dates gives the impression of a clear temporal orientation in the past related to the present. Dates mark time in a way that events in different places can be placed in the same progression. Trinidadians commonly employ this temporal strategy for structuring representations of the past, but while dates imply a steady, linear progression of time, Trinidadians' use of dates by themselves suggests that they mostly orient an event in the past when no other historical or biographical landmark is available, or dates serve as a form of redundant temporal structure. The date of one's birth, the death of a loved one, and many other significant life events are often marked by dates. An example of this is "Mr. Raj's" discussion of losing and regaining his driver's permit:

> I loss my permit. That was in '59. I can't remember the month, but I could remember the year. It was in 1959 ... Well in '61, in '61, because I could remember when I get back the permit, 18 months did pass already. So that was in '61 when the magistrate grant me back and did give me the letter.

Needles [a pseudonym], a tailor in his mid-30s, uses years in a similar way, including the recurring, redundant reference to a year:

> I leave the insurance work in 1980. It wasn't too long that I leave the work and I get the accident. So I leave the insurance work in 1980 and not too long after I get the accident—soon after Carnival, in 1980.

Even though dates imply a sequential, linear time, remembered events are not spread evenly through time. Moreover, the use of dates should not be naively interpreted as forming an accurate chronology. Figure 4.1 graphically represents, according to date, all the events recalled by research subjects in the nondirected interviews. This figure, while an amalgam mentioned during interviews with 27 people, suggests an intersubjective, cultural pattern.

This clustering becomes even more significant when viewed in terms of other means of temporally structuring the past.

SIGNIFICANT HISTORICAL EVENTS OR PERIODS

Figure 4.2 shows that the clustering of the 177 events shown in Figure 4.1 is, in many cases, a clustering around significant historical periods, such as World War II, or specific events such as general elections. While the clustering around events of historical significance is very clear, this is not because the historically significant events are always the ones recalled. The

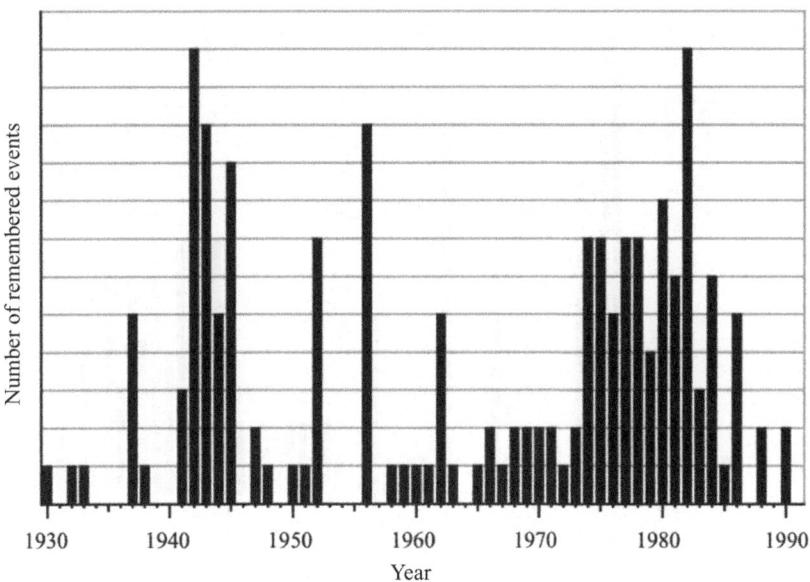

Fig. 4.1 Number of remembered events by year

local economic impact of events such as the construction of the American military bases or the political patronage that emerged with elections and party politics indirectly led to such moments of personal significance as buying land or purchasing one's first car. The representation of the past through reference to such periods is a means of temporal representation that differs from simple chronology, and also a representation that clearly manifests clustering of remembered events. One significant difference between these two temporal structures is that a connection of a memory to a major event or period of national or international significance links the speaker to history in a way that is far more direct than referring to a date. A statement as simple as "After the base, I came up here" contains a temporal reference to the American base in World War II and a sequence of events linking the construction of the base to the speaker's autobiography.

"Tantie Leslie's" discussion of her experiences during World War II serves as one of many examples of how historical events are used to structure representations of the past and mediate between global history, local events, and personal memories. In 1998, Tantie Leslie was reaping

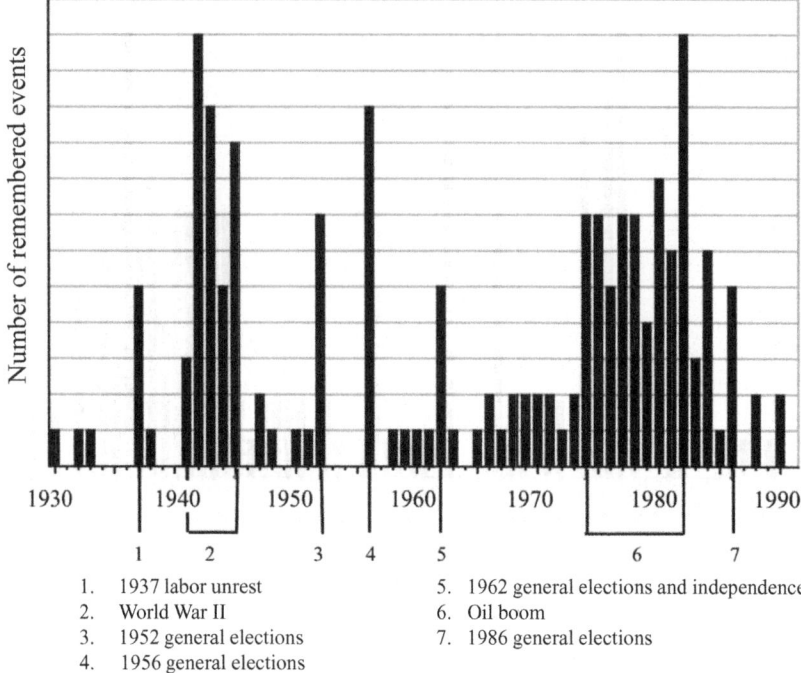

Fig. 4.2 Number of remembered events by year related to significant historical events

1. 1937 labor unrest
2. World War II
3. 1952 general elections
4. 1956 general elections
5. 1962 general elections and independence
6. Oil boom
7. 1986 general elections

the rewards of a long life of hard work. Sitting on a chair in her gallery, shaded by the huge mango tree in front of her house, she recounted for me parts of her life story. This day, it was a poignant description of her life during the American occupation—World War II. She was newly married then, and she described rising at two o'clock in the morning to cook a lunch for her husband to take to work. Her husband worked building Waller Field, an American Army Air Force base in central Trinidad. His commute involved a 10-mile bicycle ride to a town near the base from where the workers were met by a truck that then carried them past the gates and security checkpoints to work. If a worker missed the truck, he missed the day of work and its wages. The wages were very good, so even though waking up early and cooking by the light of the stove and a kerosene lamp was a hardship for Tantie Leslie, the wages her husband earned were worth it. Still, she recalled always being tired, but it was not just the early-morning cooking that put a strain on her. After her husband left, she worked on a former cocoa estate growing cassava and dasheen—two root crops—as part of the colony's "Grow More Food" campaign; in addition to that, she worked as a water carrier for government crews maintaining and building roads. The Grow More Food campaign was orchestrated by the Anglo-American Commission and implemented by the colonial government in response to Trinidad's long, historical dependency on imported food—a food supply jeopardized by U-boats and the commitment of ships to carrying war supplies instead of food. So, while rice and flour were strictly rationed, and sometimes only available at the post exchange (PX) on the bases, large expanses of land had their cocoa and coffee trees cut down in order to grow tubers. This was mostly on plantations that had gone bankrupt in the 1930s due to depressed global markets for cocoa and an infestation of witches'-broom disease—a crop disease that can only be controlled through cutting the infected branches—and this required a workforce that many plantations could not afford. Building roads was also in response to the American military's desire to move about the island with ease. Fifty years after the events, this intersection of global forces, local events, and personal memories still provided an important temporal anchor to Tantie Leslie's representation of her past. Interestingly, her first response to the extremely open-ended question "Tell me about 'long time'" temporally positioned herself in relationship to her work, her husband's work, and the period of the American bases.

K: Tell me about "long time."
T: Well, hmm, long time things hard. Cocoa and coffee were cheap. You worked for five cents per day, but then the Americans come. When did they come? 1937, '39, '40—around then. Everyone work on the base. The Americans pay double the salary that we were getting in agriculture. The government increased pay at the time—women getting half of what the men were paid. My husband work for several companies, but I don't know what they are.

As was the case in many interviews, the years she gave were not accurate—construction on the bases began in 1941. The use of dates is thus a rhetorical device as much if not more than a chronological statement.

Her oldest son also anchors his personal narrative in World War II:

> Well, normally, when I was growing up from a child, I did not grow with one parent 'cause my mother and father separated since I was about, say, one year old. Well, when I come to know myself, then, let us say approximately three years, I could remember things from then—I was with my mother. The base was in existence, the American base.

Her youngest son, Cyril, grew up at a different time. Born in the 1950s, he came of age during a period known as the Oil Boom. Trinidad and Tobago's economy relies heavily on fossil fuel revenues. With the oil embargo that Middle East oil producers placed on the United States in the early 1970s, countries such as Trinidad and Tobago benefited from a huge increase in oil revenues. Between 1974 and 1982, these revenues greatly enhanced the government's ability to expand its employment programs and subsidies for industry. This had a striking effect on Trinidad's economy with a massive expansion in manufacturing and construction. Cyril's personal narratives focus on this time of plenty—particularly on his job building cars at the Amalgamated factory, in fact not far from where his father worked at Waller Field.

Stage in the Life Cycle

In Figure 4.2, the clusterings of remembered events clearly correspond to events and periods of historical significance. I now want to take this material and exercise another transformation—this time away from dates and history and toward biography. Figure 4.3 displays the relationship of the remembered event to how old the subject was when he or she experienced the event. This figure contains 180 events, and sometimes it was not possible to corroborate the age of the subject at the time of the event by additional information. While remembered events cluster very clearly around important years and historical events, they also cluster very clearly around age. The sample of 27 people from whom I gathered these memories included men and women with the youngest person being 22 years old and the oldest being 87. The average age in the sample was 52, and the median age was 48. A clustering of events remembered from when one was young is not surprising since these are the ages all in the sample have lived through. Even so, the clustering is clearly in the late teens to early 20s. Also, despite only half the sample being above the age of 48, there is another clustering in the late 50s. When the number of memories is weighted, as in Figure 4.4, the clustering becomes even clearer around the 20s and late 50s. There are significant gaps in childhood and the period from the mid-30s until the mid-50s. This clustering calls for some explanation. Just as the clustering of remembered events

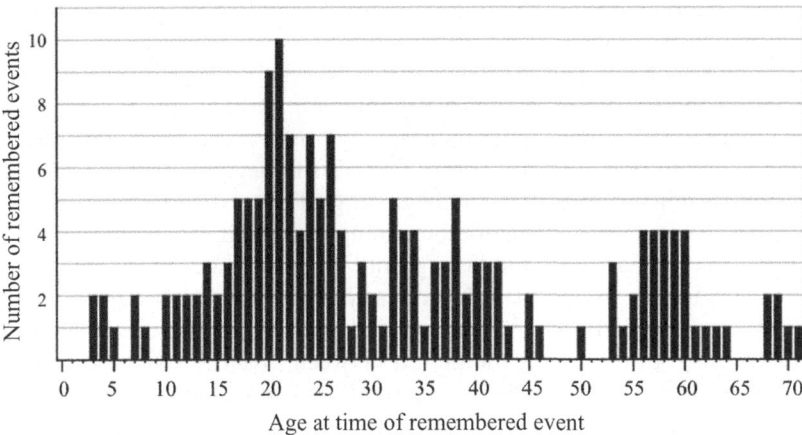

Fig. 4.3 Age at time of remembered event

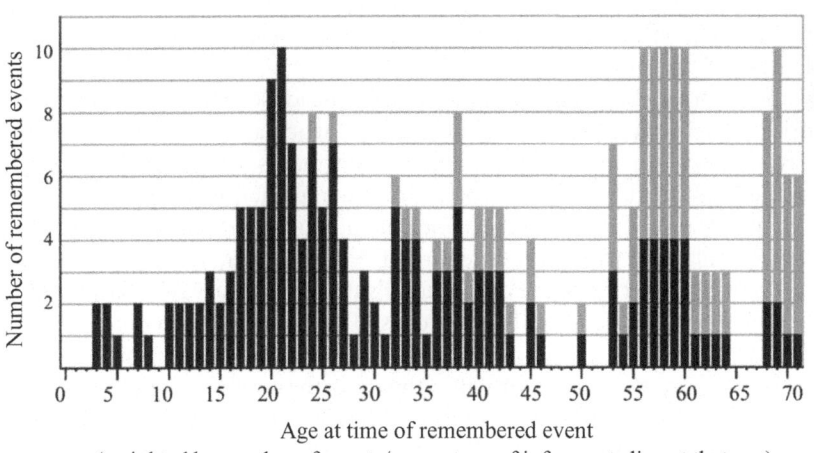

Fig. 4.4 Weighted age at time of remembered event

around years was clarified by reference to historical events, the clustering of remembered events around particular ages is clarified by reference to culturally recognized life stages—both Halbwachs (1992) and Bartlett (1932) refer to adolescence as one such stage.

Trinidadians discursively clearly identify four life stages and imply a fifth. The four they identify are "babies," "children," "youth," and "old heads." Babies are prelinguistic and not toilet trained. Children are associated with the ethnopsychological concept "coming to know themselves," as well as going to school. The major event that marks their transition to youth is taking the Common Entrance Examination. Passing this test allowed one to attend secondary school for free, and since there are no secondary schools in Anamat, this test marked either the beginning of the young person leaving the village or the beginning of the young person's working. "Youth" refers to the period beginning with the independence gained during the teenage years and lasting until an individual settles down with a family—usually in one's late 20s or early 30s. Old heads are the elderly, and it is typically those who have retired from work who are labeled as "old heads." Interestingly, the period between "youth" and "old head," in which people obtain jobs, get married, and raise children, is not discursively marked. If recollected events are graphed not only according to age but to life stages, one notes that there is a strong clustering around the period of youth and around the period of old head (see Figure 4.5). Not only is the stage of life between youth and old head not discursively marked, but it also seems remarkably devoid of remembered events, as Figure 4.5 shows.

Both those who are middle-aged and old heads seem to have their memories cluster around the period of youth—particularly obtaining their first paying employment. Indeed, for all my informants, the first job was a significant landmark in their personal stories. Importantly, though, they did not all obtain their jobs at the same chronological age, but viewed themselves as all being within the category of "youth." "Mr. Teacher," who had just retired from one of the community's two elementary schools in 1989, had first been a teacher in his midteens. This contrasts with Sinclair, who obtained his job at the Amalgamated automobile factory in his late 20s. In both cases, the timing of when they got these jobs was influenced by larger political and economic forces. In the case of Mr. Teacher, he became a teacher during World War II, a period when all the men older than he were working for high wages constructing the base or in the eradication of malaria. Sinclair received his job during the Oil Boom, a period during

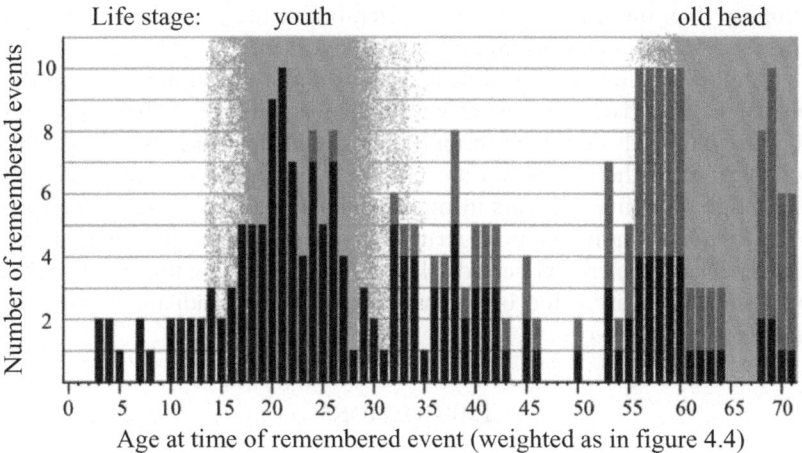

Fig. 4.5 Weighted age at time of remembered event related to culturally marked life stage

the 1970s when industry rapidly expanded. Significantly, then, transitions from one stage of the life cycle to another intersect with events of historical significance in important ways. This was also felt by many of the old heads who were forced into early retirement, and hence early old head status, when the economy contracted in the 1980s.

Trinidadians' temporal structures of the past often emphasize a period of young adulthood—these temporal structures are powerful enough to define groups. Those who worked on the military bases in World War II have that as a significant memory, and it is a memory that defines their lives. More subtly, it is a collective memory that defines their peer group— the men who worked on the base were not the same age, but come to be represented as if they were. This was powerfully demonstrated to me at a "cook" to which I was invited. Gathered around a pot on an open fire in which fish broth was simmering, the men at this cook reminisced about their days working at the Amalgamated automobile factory, and addressed each other as "cousin" rather than by name. They were different ages, the oldest being almost ten years senior to the youngest. This age difference was meaningless in the face of shared stories, experiences, and songs. The cook and I were the only ones left out of the exercise in nostalgia. We did not share the experience of Amalgamated. I was not in Trinidad at the time these men worked there, and the "cook," although only slightly younger than his relative who was a member of the group, had come of age after Amalgamated started to retrench workers.

Memory organizes their personhood around a significant event in the history of Trinidad—they were youth then regardless of their age, and they are old heads now even though their ages differ. Likewise, those who participated in the Oil Boom are not the same age as one another—an age range spanning 12 years in my sample. It is as if the critical periods exert influence on the temporal structures of narratives by pulling personal events toward the critical periods. These relationships established by temporal structures are often between historical events, individual stages in the life cycle, and work.

Implications

The data show that memories cluster in two ways: around events of historical significance and around culturally recognized life stages and transitions. Time is not mere chronology but a weaving of a seemingly steady progression of time in the form of dates into historical periods and life stages of significance. In fact, chronological dates are deceptive—they are sometimes incorrect and their use is to give veracity to the public presentation of a memory. This sense of time is more about the intersection of different concerns than a single, uniform timescale. In this way, it parallels the idea of intersecting methods of reckoning time that I developed in the previous two chapters. Time ceases to be homogeneous and empty. Simultaneity is a means of linking one's story to times that are intersubjectively significant, not a means of objectively demonstrating that events happen at the same chronometric moment. Chronology does not sufficiently capture this, for it is one of several strategies used for the generation of temporal landmarks. Historical significance turns out to be based on events that substantially alter local labor relations, and life stages seem to be importantly defined by changing labor roles—a pattern that resonates with the ethnographic literature on the Caribbean, such as Mintz's classic *Worker in the Cane* (1974).

With regard to the use of temporal landmarks to structure memory and history in Trinidad, it is the process by which temporal landmarks are chosen and used to shape and orient narratives of the past that is important. In Trinidad, temporal landmarks are based on events that involved significant changes in social relationships, and as a consequence of this, these are events that remain intersubjectively significant in efforts of being a social person with a past that is comprehendible to others. Furthermore, when the temporal landmarks employed for organizing the past in other societies are examined, whether it be the rich and growing literature on Madagascar (Bloch 1986, 1998; Cole 2001), Hoskins' study of the Kodi of Indonesia (1997), or even Evans-Pritchard's discussion of structural time among the Nuer (1940), it becomes clear that temporal landmarks elsewhere can be defined by kinship or cycles of ritual. This does not imply that labor is not intersubjectively significant elsewhere, but because kinship and ritual are not used to define temporal landmarks in Trinidad, it does suggest that there is cultural variability in the choice of landmarks. Thus, memory efficiency and function might only be sufficient to explain why temporal landmarks are used, but an appeal to individual memory efficiency and function cannot explain why events become temporal landmarks, why landmarks cluster, or why there is cultural variability in patterns of choosing temporal landmarks. To answer these questions, one must explore the ongoing flow of social discussions and internal dialogues that tie representations of the past to identities, ideologies, and political economy. One must also give up the assumption of temporal uniformity that organizes chronology in favor of the fits, starts, and lacunae of narrative temporalities.

The role of the past in the conceptualization of the nation has received a great deal of attention. From Hobsbawm and Ranger's idea of the invention of tradition (1983) has arisen an academic perspective that looks at time from the perspective of the construction of history for nationalistic purposes combined with the imagination of the future. For instance, Bhabha offers a concept of "double time" in which a mythical view of the past is used to imagine a great future (1994, 144–145). In the Caribbean, this creates profound difficulties. The colonial past of exploitation for the enrichment of Europe and its North American colonies is not a grand mythical age, and when coupled with the tale that the indigenous population disappeared—a myth which is actually false, but nonetheless powerful (Forte 2005)—there is no previous time to which West Indian societies can refer as their golden age. Fanon bitterly wrestles with this fact: "The black schoolboy in the Antilles, who in his lessons is forever talking about 'our ancestors, the Gauls,' identifies himself with the explorer" (1967, 147). For Fanon, building as he does a synthesis of Sartre, Lacan, and Kojève with his experience as a Black Martiniquan in France (Bulhan 1985), such identification is the source of a psychological and social struggle, if not confusion. Counter to Bhabha's idea of a double time, Fanon suggests, "The problem considered here is one of time. Those Negros and white men will be disalienated who refuse to let themselves be sealed away in the materialized Tower of the Past" (1967, 226).

Edensor complains, "[A]ccounts of temporal constitution of national identities continue to focus upon common traditions, myths of shared descent, and the linking of historical and future narratives, reflecting a wider obsession with official, historical, elite constructions of national identity" (2006, 527). Such shared myths have received much criticism. But as with clock time and labor-time, the temporal assumptions of European-derived historical narrative genres elide other ways of constituting the past. It is not just the content of history or the perspective of the tellers of history that is at issue, but the representation of time that shapes the telling. Ever since the creation of universalizing chronologies in the Renaissance (see Wilcox 1987), history has been constructed out of a relationship between narrative time and absolute chronology. Postcolonial criticisms of history have raised the constitution of narrative (see Scott 2004), but take chronology for granted. Chronology is a contrived and expert cultural rhetorical knowledge.

Representations of history commonly adopt a chronological rhetoric to anchor narratives and yet violate the qualities of sequence and simultaneity that are seemingly essential in chronology. Historical narratives commonly skip lacunae in the course of events and privilege some simultaneous events over others for the sake of the story. Contrary to the implication of homogeneous, empty time that dates are containers to be filled with events, dates become islands in time—landmarks that anchor events even as the representation condenses the sense of time in the narrative.

The discussion of temporal landmarks that organize recollections of the past makes the same basic point made in the previous chapters—homogeneous, uniform representations of time distort understanding. Just as chronobiology, psychophysics, and work do not conform to homogeneous, uniform timekeeping, yet become represented through that lens, so too does the past tend to get represented within a framework of homogeneous, uniform units of time. Centuries are not meaningful boundaries in social processes or memories, yet they become the framework by which processes and memories get discussed. That does not mean that such units are not useful. They are of use in thinking about duration between events or charting the relative simultaneity of events, but there are several pitfalls in their use in constructing narratives of the past. First, they create divisions between time units that divide continuous social processes. More important for the understanding of how memory influences actions in the present is the way in which those actions lead to a construction of events that, while chronologically accurate, does not match the ideas of the past that shape social processes. The pull temporal landmarks exert on autobiographical events leads, strictly speaking, to historical inaccuracies, but it is not a chronologically structured academic history that drives how memories affect the present. The past in the present is not conditioned by uniform chronology, but by intersubjectivity.

CHAPTER 5

Tensions of the Times: Homochronism versus Narratives of Postcolonialism

Édouard Glissant writes, "To confront time is … for us to deny its linear structure" (1989, 145). In Euro-modernity, a linear concept of time that employs the metaphor of time units as containers to be filled is a critical component of planning and how the future is imagined. In contrast, utopian thinking utilizes several different temporalities. In her discussion of utopia, Levitas (2013) points out the contrast between those who see utopia as a goal versus those who see utopia as a method. The dominant temporality for the former is to see utopia as timeless and temporally disconnected from the present. The dominant temporality of utopia as method is to emphasize change and transformation of present conditions by means of using the imagining of utopia to critique the present. The three models of linearly arranged containers versus timeless separation versus kairotic change are, in fact, three alternatives for the temporal organization of conceptualizing the future. Just as with the past, there is a tension between thinking that emphasizes chronology and thinking that emphasizes values.

What I am discussing here is not the future, per se, but the temporalities that are used to think about the future. As with chronobiology, the representation of labor, and the use of chronology to structure the past, there is a temptation to rely on uniform durations and homochronicity to structure the future. Yet, there is a growing literature to suggest that alternative temporalities must be considered when studying how people imagine the future (Adam 2010; Crapanzano 2007; Lazar 2014; Miyazaki 2003; Ringel 2014).

The anthropological approaches to hope seem to imply that hope is outside of the temporality of linearity (Miyazaki 2004; Ringel 2014), or that hope even consists of imagining a rupture from the present (Crapanzano 2003; Nugent 2012). This disconnection is a product of privileging the metaphors of linearity and time as consisting of containers to be filled. Other temporalities allow for a connection between hope and the present without the near future being evacuated. For instance, Ringel (2014) discusses how ideas of hope can be tied to ideas of the future that emphasize stability. Such continuity is unappealing to a temporality that emphasizes discrete units of time extending into the future. Containers of time that are filled with the same routine seem unappealing to temporalities of modernity, as Sherman's discussion of the emergence of the genre of English diaries shows (1996). He describes how diaries changed from emphasizing repetition to developing narratives, and toward seeing uniform durations as something to be filled. This applies to the future, as well. Contrary to Guyer (2007), Christians looking forward to the apocalypse have not vacated the near future, because their hope is that the apocalypse is near. The near future only appears to be vacated through the lens of the dominant, secular temporality.

Opitz and Tellmann suggest that the futuricity associated with economic temporality "disrupts and disfigures legal temporality"—particularly legal protections (2015, 108). Their insight suggests that the dominant temporality associated with time as linear and consisting of uniform containers is disruptive to alternatives, including temporalities of hope. Maybe this is why Miyazaki remarks that the "temporal orientation" of her analysis was "incongruous" with the ritual practices associated with hope that she studied (2004, 10–11). Her observation is grounded in the philosophy of Ernst Bloch, who argued that the approach to temporality inherent in Enlightenment European philosophy was not capable of imagining the future. Instead, he urged an anticipatory consciousness that perceives the immanent potentialities of the present (1986). The clocklike temporality that defines time as a progression of uniform containers limits imagination. It already provides an inescapable and nonnegotiable structure for the future.

Miyazaki sees a contrast between the strategy of moving away from clocklike temporalities and the strategy of apprehending a now in which hope is immanent (2004, 19–22), but these are not exclusive strategies, and as I shall argue and have been arguing, embracing alternative temporalities is necessary for seeing biology, labor, and the past in meaningful ways. This applies to the future and to hope, as well. As Miyazaki argues, hope prompts a "radical temporal reorientation of knowledge to the future" (2004, 7), and I would suggest that this reorientation defies the dominant temporality of linear, clocklike time in favor of a narrative time that weaves together the present and the future.

The conceptualization of time as a linear arrangement of time units to be filled is also closely linked to the technoscience of time metrology, the standardization of which was important for managing empires. This was true in the ancient world (Stern 2012), and remains a critical feature of global regulation and control. It is a small leap from such practices to Marcuse's argument that technical reason is ideological (1968, 223). In the form of technocratic knowledge, so crucial for ideas of "planning," the science of metrology becomes merged with the exercise of power with the metrological science, making this merger of science and power seem "less ideological" and yet all the more a threat to humanity's emancipation (Habermas 1971, 111). Or, as Virilio puts it, "This is indeed one of the unacknowledged aspects of the globalization of a real time that subverts not only the real space of the geography of the globe, but also our relationship to time that is really *present*" (2010, 92).

The uniform, linear temporality that dominates current time metrology is one temporality among many, however (see Greenhouse 1996). As Stewart's work on Naxos (2012) and Kermode's work on narrative (2000) suggest, a challenge to thinking about the temporalities that shape the imagination of the future is the consideration of how these temporalities are related to the temporalities that shape the past or the present. Temporalities are like Deleuze and Guattari's (1987) idea of the rhizome: They proliferate and imbed themselves into all sorts of thought and action. To divide discourse into past/history, present, and future suppresses the ability to see how temporalities spread across these conceptual divides in a rhizomatic fashion.

This can be inferred from an intertextual reading of Guyer's "Prophecy and the Near Future" (Guyer 2007) with Bachelard's *The Dialectic of Duration* (2000) and Henry Fielding's *Tom Jones* (1963). Guyer argues that in both evangelical Christian and macroeconomic temporalities, the near future is neglected—"evacuated," as she says (2007). This absence of an explicitly imagined near future creates an episodic representation of the future in which important events are not directly connected to the present. Evacuated periods are not limited to macroeconomic and Christian conceptions of the future, however. They are a common feature of narratives.

For instance, Henry Fielding mused about a type of evacuated period in *Tom Jones*:

> We intend ... rather to pursue the method of those writers, who profess to disclose the revolutions of countries, than to imitate the painful and voluminous historian, who, to preserve the regularity of his series, thinks himself obliged to fill up as much paper with the detail of months and years in which nothing remarkable happened, as he employs upon those notable areas when the greatest scenes have been transacted on the human stage ...
> Now it is our purpose, in the ensuing pages, to pursue a contrary method. When any extraordinary scene presents itself (as we trust will often be the case), we shall spare no pains nor paper to open it at large to our reader; but if whole years should pass without producing anything worthy his notice, we shall not be afraid of a chasm in our history; but shall hasten on to matters of consequence, and leave such periods of time totally unobserved. (Fielding 1963, 54)

These omissions are what Bachelard would call lacunae in time—periods in which nothing of note happens (2000, 19).

Fielding's approach to narrative defies uniform, homogenous, sequential temporalities in favor of episodes; Bachelard defies similar temporalities in his study of consciousness. In contrast, according to Guyer, economics and apocalyptic Christianity use such temporalities to demonstrate that the near future is evacuated.

In commenting on Guyer's discussion of the evacuation of the near future, Crapanzano (2007) chooses to develop the theme of multiple and conflicting temporalities rather than to dwell upon Guyer's claim of the similarity between macroeconomic and Christian concepts of the future. Multiple temporalities are well documented for the past and present, so it makes sense to consider multiple, conflicting temporalities as an aspect of how the future is imagined. Rather than focus narrowly on the future, one should link the future to the multiple temporalities in the present and the multiple chronotopes and narrative structures that link the past, the present, and the future. The future is enriched by narrative, and narrative depends on referencing the past.

The discussion of temporality intersects with a more pervasive concern of how the future is represented in relationship to the present and the past. Is it connected to and caused by the present and the past, or is it conceptually separate? The debate about climate change involves the relationships of the future to the present and the past. Those who argue against humanity's role in climate change point to the climates of the past and argue that whatever change that is observed is "natural" and fits in with long-term climate cycles. The temporality latent in this position emphasizes repetition and assumes that patterns of the past are always predictors of the future. For instance, at the beginning of *Unstoppable Global Warming*, Singer and Avery state, "Through at least the last million years, a moderate 1,500-year warm-cold cycle has been superimposed over the longer, stronger Ice Ages and warm interglacials" (2008, xvii). In contrast, those arguing that anthropogenic climate change is a crisis suggest that there has been a transformative change in the climate so that the old patterns no longer apply. As Al Gore puts it in the introduction to *An Inconvenient Truth*, "The relationship between human civilization and Earth has been utterly transformed by a combination of factors, including the population explosion, the technological revolution, and the willingness to ignore the future consequences of our present actions" (2006, 8). The contrast between "a moderate cycle" and an "utter transformation" is stark. It reveals completely different ideas of the relationship of the past, the present, and the future. In one, the present and the future repeat the past; in the other, the present is a pivotal moment in the future unfolding on principles different from those that shaped the past. Both involve assumptions about immanence (and sometimes imminence). In the one position, one can find uniformitarianism—the geologic principle that the processes at work in the present were at work in the past and will be at work in the future. In the other, what is immanent in the present is a tipping point in which future processes will not be simple repetitions of the past. The projection of cycles into the future is dependent on an interpretative lens that projects those climate cycles into the past. The projection of radical change and crisis in the future is dependent on an interpretive lens that human activity has Earth-altering consequences so that past patterns no longer apply. The temporality of cycles removes humans as having a significant influence over Earth; the temporality of climate change emphasizes the transformative, and sometimes negative, consequences of human activity.

These different temporalities drive different policy positions. If humans have little or no influence in comparison to large-scale cycles, then humans should not sacrifice economic growth to combat something that humans would be ineffective at combating because it is "natural." On the other hand, if Earth is the way it currently is predominantly because of human activity, then human agency can shape the future. As in any dialectic, one should not be surprised when a middle position emerges.

Such is the case with the position of the U.S. Department of Defense. Since both sides in the debate accept climate change, the Department of Defense accepts climate change and has publicly crafted policies to adapt to it: "The third National Climate Assessment notes that certain types of weather events have become more frequent and/or intense … These climate-related effects are already being observed at installations throughout the U.S. and overseas and affect many of the Department's activities and decisions related to future operating environments, military readiness, stationing, environmental compliance and stewardship and infrastructure planning and maintenance" (2014, 2). In the Department of Defense's position, there is a connection between the present and the future such that the seeds of the future are in the present.

Immanence and imminence are particularly influential in policy discussions to guide present actions in relationship to anticipated conditions. For instance, the U.S. Department of Defense is very concerned about climate change in ways that reflect this combination of immanence and imminence. In contrast to the climate-change debate, Guyer describes the "evaporation of the near future in theory and public representations" in macroeconomics and Christianity (2007, 410). She notes an absence of the immanence of the near future in contrast to the sense of immanence found in the Department of Defense's concern about climate change. She notes that in macroeconomic theory, the emphasis on growth and management technology creates future goals that leave the immediate consequences of current actions out of how the future is imagined. Likewise, in some forms of Christianity that emphasize the Second Coming of Jesus Christ, "the near future is evacuated, in a way that is just as disorienting and yet internally logical as its secular counterpart in economics" (2007, 414). The similar stances toward the near future in economics and Christianity seem to indicate a cultural trend.

The disjuncture between the emphasis on immanence and imminence in concerns about climate change, on the one hand, and the elision of the near future in economics and apocalyptic Christianity, on the other hand, is not a matter of one representation being incorrect, but an indication of the complexity with which the future can be imagined—the possibility of multiple, coexisting images of the future, as Crapanzano (2007) suggested. Yet, even if one acknowledges that multiple imagined futures exist, there are other elements that are not found in the discourses of climate change, macroeconomics, or the apocalypse, namely, cultural reproduction and conservatism.

Cultural conservatism is a feature of temporalities that is often unstated. Climate-change arguments are explicitly struggling against persistent habits that release climate-altering gases into the atmosphere. Evangelical Christians anticipate Jesus' return but can describe their weekly schedules of Bible study and church services and predict those cycles extending into the distant future. Macroeconomics emphasizes the timeless consistency of theoretical principles while positing a future that is different from the present. The near future is not empty, but consists of structures of action that ensure continuities from the present.

The epistemological and cognitive hold of uniform timekeeping over thought is not a matter of change, but is instead a matter of cultural conservatism enabled by the reproduction of cultural ideas of time by clocks and calendars. As cognitive tools that think for their users, clocks and calendars are powerful means of cultural reproduction (Birth 2012). This is in contrast to how technology is normally considered. Technology is often portrayed as an agent of change (Mumford 1963; McLuhan 1994). Here, instead, I am emphasizing how certain technologies reproduce structural arrangements of power. As such, this reproduction is extended into the future, and noting where such extension is evident and where there are ruptures is important in understanding how the future is imagined. As is the case with studies of psychophysics, chronobiology, the representation of labor, and the chronological conception of the past, lived multiple temporalities challenge the homochronic logic that much scholarship and public policy seeks to impose.

Trinidadian Futures

It was a hot afternoon in August of 1990, and a group of people were drinking beer at a place known locally as "the Junction." The beer was warm because the village had been without electricity since shortly after the attempted coup d'état of a few days before. The warm beer was provided by a young man who had just lost his job due to the ongoing recession and who was using his severance pay to buy rounds for everyone present, whether they wanted to drink or not. As the alcohol took effect, he said, "I have no future, so I goin' to get a gun and shoot some fuckin' police and go out famous. I want to be front page on the *Express* [one of the two daily newspapers] so that everybody know my name." Thankfully, he did not act on his desire—he has settled down and has a steady job.

Despite his life not following the path he predicted in 1990, his words are still a vivid memory not adequately captured by the text of my field notes.

His talk that day contains a dense web of temporalities and images about the future. His feeling of having no personal future contributed to his plan of being memorialized on the front page of a daily newspaper—a medium Anderson associates with chronological dates and homogeneous, empty time (2006, 22ff.). The continuity of the newspaper and its dates from the present extending into the future made it a reliable structure for the future with which to fantasize about a future action. Implicit in this man's discussion was a parallel between the recent downturn in his personal finances and the bleak economic prospects of Trinidad's economy at that moment. This was not a continuity but an implicit rupture between a past of gainful employment and a bleak future of joblessness in a recession. The recession had devastated the local economy and the attempted coup violated the ethos of Trinidad as a peaceful place. While this young man was particularly nihilistic in August of 1990, many of his peers entertained fantasies of escape—particularly through emigration. All were crafting narratives of the future that were seeking endings that were consistent with the lived middle/present—what Kermode calls a "sense of an ending" (2000).

The young man threatening to kill police on that August day wove together many concerns that were on others' minds even though his sense of an ending was different from that of most people drinking beer with him. He brought together the themes of the attempted coup and the recession and allowed multiple temporalities to coexist: the continuity of the newspaper that would document the ending of his own life. The moment he created was poignant, but it was not disconnected from my other fieldwork experiences and discussions. Since concepts of time are embedded in everyday discourse and are rarely explicitly addressed, during my fieldwork I listened a great deal to discussions about the present state of affairs in Trinidad in relationship to the past and the future to learn embedded temporalities. During my stay, two crises captured the imagination of those with whom I lived and worked. One involved the deteriorating economy—a situation that produced rampant retrenchments, the elimination of cost-of-living allowances, the devaluation of currency, and the imposition of a value-added tax. The other crisis involved the attempted coup d'état of 1990. The discussions of these crises displayed two different temporal logics: discourse about economic crisis adopted homogeneous temporalities; discourse about the coup attempt adopted narrative temporalities. The differences between these two temporalities are that narrative temporalities involve a story arc—a development of suspense leading to a conclusion; and that they demand no temporal connection between the present and the future other than the narrative arc—the future is imagined as episodic and populated by temporal landmarks with unarticulated lacunae in between. This is very similar to the discussion in the last chapter in which the past was conceived in terms of landmarks that seemed to attract events toward themselves but to leave large chronological gaps. In contrast, homogeneous temporalities treat all time as alike—there is no greater significance given to one moment versus another.

The recession and the attempted coup d'état involved creating temporalities. The distribution of different temporal logics over various narratives reveals conflicted temporalities in how the future is imagined. Such multiple temporalities challenge the idea that homogeneous empty time is the dominant temporality of modernity (Taylor 2007) as well as a necessary temporality for the imagination of the nation (Anderson 2006). Instead, they suggest a contrapuntal relationship of temporal logics in which different contexts and situations influence different imaginations of the future.

The co-emergence of the modern forms of narrative and homogeneous empty time poses a dilemma—they coexist and are mutually exclusive. This dilemma is usually elided by placing narratives within homogeneous empty time without noting that the temporalities of narratives undermine the image of homogeneous time. Bhabha senses some of this tension: "If it is the time of the people's anonymity it is also the space of the nation's anomie" (1994, 159). Homogeneous empty time is tied to a loss of meaning and thus to a loss of the narrative imagination of the future.

Similar to the ethnographic material of the last chapter, Caribbean writers also express the tension between multiple temporalities and homochronicity. For instance, Glissant writes, "Our quest for the dimension of time will therefore be neither harmonious nor linear. Its advance will be marked by a polyphony of dramatic shocks, at the level of the conscious as well as the unconscious, between incongruous phenomena or 'episodes' so disparate that no link can be discerned" (1989, 106). Wilson Harris contrasts the "inner clock" with the "static clock that crushes all into the time of conquest" (1999, 179). To grapple with modern temporalities, one must understand that the reproduction of the conflict between homogenous time and prefigured time has been a feature of modernity since at least the Enlightenment, and consequently, the reproduction of this conflict has been a feature of colonial and then postcolonial conditions.

An Economic Crisis

In Trinidad and Tobago, economic discourses invoke multiple temporalities. The state's budgets and planning processes coexisted uneasily with the IMF's emphasis on debt servicing and economic restructuring. As Simone Abram has pointed out (2014), even when planning processes seem to be using the same temporality, there can be conflicts between their rhythms and schedules. Periodically coincident with these ideas were parliamentary elections—if the plans and promises of the majority party were overshadowed by the pain of the IMF's required policies, its election prospects were bleak. This was particularly true because electioneering in Trinidad involves expenditures on temporary work projects in order to cultivate votes, and such public expenditures were exactly the sort of practices the IMF requirements were meant to squelch.

Trinidadians' conversations with me about the present and future of the nation often focused on if and when the recession would end, and also often involved a comparison between the recession and "when things bright" during the Oil Boom. These were not images of the future disconnected from the past. On the contrary, they were framed in reference to the past—particularly the Oil Boom and the political dominance of Eric Williams and the People's National Movement (PNM).

During the Oil Boom of the 1970s, the state had abundant capital to invest and to employ those people not working in the private sector, but as petroleum revenues decreased in the early 1980s, these fiscal commitments turned what was a seemingly endless supply of capital into debt; within a decade, Trinidad and Tobago turned from a lending nation to a nation seeking debt restructuring from the IMF.

The steward of the petroleum windfall and the resulting state spending that led to indebtedness was Eric Williams and his party, the PNM, which held a majority in parliament from 1962 until 1986. The history of the PNM and Williams was closely tied with the political party taking shape around the "University of Woodford Square," during the mid-1950s, in which Williams was a prominent speaker. From these political gatherings in a park in Port of Spain, a movement gathered steam to take advantage of the granting of universal suffrage in 1946 and the move toward complete independence. Williams quickly became the leader of this movement, and in the 1956 elections, the PNM was able to win enough seats in the Legislative Council to convince Great Britain to grant the PNM the power to form a government, rather than merely to advise the colonial governor. In 1962, Trinidad and Tobago achieved full independence, still with Williams at the helm. Williams maintained tight control over the PNM and served as the head of state continuously from 1956 until his death in 1981, presiding over many parliamentary electoral victories.

Williams' legacy is contested, and embedded within the different perspectives on Williams are different temporalities linking the past, the present, and the future. For instance, one retired Indian taxi driver offered me a detailed critical history of decolonization. Like some of his age and ethnic background, he lamented the end of colonialism. In the context of discussing Trinidad's recent history, he launched into a critique of the policy of import substitution, a policy which has since become a standard component of models of economic development and restructuring:

> Before, when white people ruled this country, things were better. We had solid, powerful American car. We had quality tire. When you wanted flour, you could go to the store and choose which kind of flour you wanted. Today we have expensive Japanese car, and our tires don't wear well—in three months they blow. We now have a flour mill, but we grow no wheat! We used to get high-quality things from the States, but now we make things here, and they are no good. After the white people ruled, Gomes and his friends were in power. They left the treasury with 21 million dollars in it. But then Williams came and depended on the petrol dollar. Williams did not see how the petrol dollar would rise and fall. He spent all that 21 million! Now we are in debt—we have to borrow money.

The temporal structure of this narrative was of a rupture between the colonial past and the postcolonial future. In it are changing places of influence—the colonial past indicated trade with the United States and quality products that endured (itself a material temporality), whereas the postcolonial present is associated with Japan and products that do not last. The process of separation between the past and the present is in the hands of two characters: Gomes, who was the political leader before Trinidad and Tobago obtained internal self-rule, and Williams. Williams is portrayed as spending a government surplus left from colonialism and going into debt. All of this is connected to the unpredictable cycles of oil revenues. The combined temporalities in this story, then, include both decline, as the result of a movement from commodities that lasted to commodities that were ephemeral, and unpredictable cycles. In addition to the narrative of decline, a striking similarity between narratives of this sort was the role played by Williams. The narratives strip him of any ability to anticipate the future—"Williams did not see how the petrol dollar would rise and fall."

Those who celebrated Williams' leadership and who articulated a list of significant accomplishments also portrayed him as unable to cope with changing global economic conditions—an African government worker told me in a narrative that praised Williams' accomplishments but still echoed the elderly Indian man's narrative about Williams' inability to foresee the future:

> There was some improvement after independence. Some big buildings were built; a steel plant was built; and there was drilling for oil at sea. They built a hospital. The Oil Boom had some effect—it made the cost of living and labor higher. There was lots of money then, and the government had 80,000 people on payroll all doing nothing. By 1982, the dollars finish. The government had to let the 80,000 people go. This was the downfall of the PNM.

In this short narrative is a mirror image of the previous one. After Williams, institutions of enduring qualities were built—big buildings, a steel plant, a hospital. But like the previous narrative, the cyclical nature of oil revenues is also emphasized. In both these narrative representations of the past there are at least two temporalities at work: things that last versus cycles that are unpredictable. Tied to unpredictability is the inability of Williams to predict.

An Indian cocoa farmer offered yet a different view, but this, too, included the idea of things that endure in relationship to change and developed similar themes about the government and oil revenues:

> In Trinidad during the boom, the government got money and supported labor. They took the idlers. The lifeline of the country is agriculture. If you do not have that, you are doomed ... We have oil, asphalt, sugar, cocoa, coffee—all bringing foreign exchange. We crying still. We have the oil industry. We have a methanol plant and a fertilizer plant in Point Lisas. China imports fertilizer from us. We have a cement factory, and yet the country is going down. What is the cause? We have been retrenching labor for years. This is to help industry grow, but it has not improved the industry.

In this statement, it is Trinidad's natural resources that endure, as well as institutions that the state constructed, like the methanol, fertilizer, and cement plants. Implicit in the narrative is, again, the government's overspending based on high oil revenues during the Oil Boom. Again, it is an interaction of permanence and cycles.

These representations contain a discussion of the past and the present, but in terms of a radical separation of the two. In the first one the narrative includes a break between the past and the present. The past is introduced with "before" or "we used to," in contrast to "but now." The second narrative signifies a break between the past and the present with "There was lots of money then ... By 1982, the dollars finish." The third distinguishes the present prefigured by the past, and the present as the speaker experiences it through the contrast of all of the government's investments, and yet the country "is going down" anyway.

These narratives also manifest the equivocal elements of political weakness and dashed post-independence hopes. The source of the lack of Williams' foresight differs between his supporters and detractors, but the sense of ending offered by both is the same—they agree that Williams' policies and actions did not prefigure the future. Williams could not determine the future. While his actions did not prefigure the future, the temporalities of permanence, ephemerality, and unpredictable cycles are all deployed when discussing the future. In fact, it is the temporalities of ephemerality and unpredictable cycles that allow the crafting of ruptures between the past, the present, and the future.

Williams' death coincided with the first signs of an economic contraction as a result of falling petroleum revenues, and his successor, George Chambers, faced this economic challenge without the academic credentials and intellectual gravitas of Williams. The narratives of Williams' leadership I was told were episodic: they commenced with the nationalist politics of the struggle for independence, skipped to the wealth and corruption of the Oil Boom of the 1970s, and culminated with Williams' death on the eve of the decline in oil revenues, and the ineptitude of his successor, George Chambers. As one elderly Creole told me, "The Doctor [Williams] take over in 1956. That time started our getting regular work and money. He led the country for 25 years. After he, Chambers take over. They say Chambers duncey." Under Chambers, the economy contracted and retrenchments began.

As a result of a lack of faith in Chambers and the PNM, the PNM lost control of Parliament in 1986 to the National Alliance for Reconstruction (NAR), a coalition party led by A.N.R. Robinson, a former Williams protégé and Minister of Finance in a PNM government. Robinson joined with other politicians who had left (or been ousted from) the PNM as well as leaders from the United Labour Front, an opposition party led by Basdeo Panday. The new NAR government quickly ascertained the financial situation of the government and sought help from the IMF. This led to the adoption of typical IMF economic restructuring policies such as cutting the public workforce, implementing a value-added tax, and infrastructure projects and tax incentives meant to encourage foreign investment.

With retrenchments, many returned to agricultural work. This was difficult in three ways. First, those who had remained in agriculture during the Oil Boom complained that the government had not invested in farming: "They keep telling the poor man 'agriculture, agriculture, agriculture'—but they do not assist in any way. They is only speaking without doing anything. The Minister of Agriculture does this." Second, because so many had abandoned farming in favor of other sorts of income, much of the profitable cocoa land had "gone abandon," meaning it had not been properly maintained by keeping the undergrowth under control and the drains dug and cleared. This meant a lot of work and a couple years of patience until the trees returned to their full productivity. An African man who had lost a factory job described working abandoned land to me as, "It come like working a fresh piece of land." The third problem, summarized by an elderly Indian cocoa farmer, was that "People get used to government work, and they became unused to cocoa work."

Agriculture, particularly if one is rehabilitating abandoned land, does not produce a steady cash flow. The shift from wage labor to agriculture created a shortage of cash that had a ripple effect on other local businesses such as rum shops, grocery stores, taxi drivers, and tailors. As I went from house to house taking a census, I found many willing commentators on the state of the nation. I would typically go through my basic demographic questions, ask members of the household if they had any questions about my work, and then encourage them to tell me whatever they thought it was important for a graduate student from the United States to hear. Often, I would stop by snackettes and rum shops to listen to people talk, and underemployment and political corruption were commonly discussed and linked. In the early days of my fieldwork, without any prompting on my part, I heard a great deal about the economy and the government's policies. The focus was often on the relationship of work and money. For instance, an Indian woman told me, "Work is hard. People are getting laid off. How are they going to feed their families this Christmas? Here in the countryside, there is plenty work, but there is not enough money." That same day, a Spanish Creole gentleman who lived on the other side of the village said:

> Labor costs about $40 a day, but cocoa is only $40 a kilo. With all the cutlassing and harvesting, you barely make enough off of your cocoa to pay for labor costs. Large land holders are selling their lands—or leaving them to go into bush. You could take loans from the government, but you can't pay these loans off. You do not make enough money. The government then repossess your land. Trinidad is in a real problem. Everything is going up, but you cannot get a decent salary. The VAT [value-added tax] will kill somebody—I don't know what a poor man will do.

These events shaped how Trinidadians conceptualized the past in relationship to the present. The latent temporalities in these conceptualizations are similar to those of the narratives about Williams. As in the discussions of Williams, the ability of the actors to determine their future is absent.

But where is the future in these discussions? The future is not mentioned. At one point, I started explicitly asking people about their futures. This produced a variety of responses, almost never tied to any narrative. One elderly Indian man said, "At this point, I cannot think about the future. At the age of 73, I don't know how much longer I will live." He would later say, "I am thinking about if I should die—how should I arrange for my children and everything else after my death." Clearly, in this man's image of the future, the only relevant event was his death. This man's sentiments were echoed bluntly by another man who lived up the road: "I don't expect anything to happen in the future except death."

In contrast to explicit temporalities of death are plans about the future. Plans forge a temporality that links the present and future. One of this man's peers in age emphasized the importance of continuing to work and save. He said:

> One would work and save a few cents. There are many different ways people would save. Some again, whether they have the ability to save, they want to have a good life and those don't care about tomorrow—no future is important for them. They going think if they have to help somebody, even if it is their own relative, they don't have any money to help. That idea—but I am not so. I say we here as humans, whether relatives or whatever it is. You can help one, help.

This man crafts a clear link between present saving and future actions—specifically, helping people. In his link between the present and the future, one works now and saves in order to help people later.

A middle-aged Creole man who was working somebody else's neglected cocoa land and splitting its revenues gave fairly concrete but limited plans about the future: "Right now I planting bananas, and I have plans for citrus, because citrus is low maintenance. Cocoa and coffee are high-maintenance trees. Citrus also has a benefit of you can sell it in the market, whereas the only market for cocoa is the dealer." While this seems to be a matter-of-fact description of his plans, it also contains a subtle break between the past and the present. The land he was cultivating contained cocoa trees that were not very productive because the owner had neglected the land. This man wanted to break from the past of cocoa and cultivate a new crop—citrus. He wanted to do this not only because the trees were easier to maintain, but also to achieve some amount of independence. Grapefruit and oranges could be sold directly in retail and wholesale markets, whereas cocoa was sold to dealers representing the government, which set standards and prices.

Other young and middle-aged men framed their future in terms of two periods—one of wage labor and the other of landownership. In 1991, petroleum exploration was done near the village, and one of the men on a short-term contract with the oil company to lead them through the local forest said, "If they find oil, I working for three years and then buying some land." Echoing this man was another who was often away from Trinidad working for a construction company. In recounting his recent experiences in Nevis, he said, " I works seven days a week, sometimes 12 hours a day. The money is good. I intend to stay in this work for the next five years, and then I will have enough money to buy a van and some land." For these two men, the near future is defined by wage labor and the distant future by independence and landownership. In a way, the emphasis on independence in the more distant future resembles that of the man engaged in sharecropping. It is a common theme—another individual who was in his 50s and about to retire from government work said, "I hoping to, well after, not to say really resignation [retirement], I looking to see if I can acquire another piece of land. Purchase another piece of land where I could have a better distribution among the children." He then assessed his prospects of doing this: "The opportunity is not yet there. There may be land available, but you know people, they can't get anything from it for themselves, but the price they would want to ask. So it would be foolish to go off and put a big set of money if it is for those prices. I am not seeing it except in a year or three."

While these visions of the future seem quite different, what is shared among them all is a close connection between the near future and the present—the near future is just like the present. The distant future is what varies. For some, the near future of sameness ends in death; for others, it ends in achieving enough money to become independent—largely from buying land. These plans for eventual independence contain a subtle temporal marker—they all refer to a finite number of years: Work in oil for three years; work in construction for five years; buy land in a year or three. They all pose a period of sameness and uniformity represented in terms of uniform units of time followed by a distant future in which they achieve their goals.

These discussions unfolded in the period leading up to and following the attempted coup d'état of 1990. The temporality of ongoing sameness into the near future followed by a rupture from that sameness is different from the narrative arc associated with incorporating the attempted coup into images of the history and future of Trinidad. In those, there is an absolute rupture between the present and the future.

An Attempted Coup d'État

In terms of underlying temporalities, comments about the attempted coup of 1990 led by Imam Yasin Abu Bakr were quite different from discussions of economic conditions. As I have previously argued (Birth 2008), the initial response was a desire to make sense of the event manifested through the ubiquitous question, "What do you think?" By the following Carnival, interpretations began to take shape, but in the hands of calypsonians, the coup narrative had elements of comedy.

In *Party Politics*, James raised the calypsonian Mighty Sparrow to the same level of national importance as Prime Minister Williams (1984, 162), and in the wake of the attempted coup, it was clear that calypsonians still did imaginative work the state could not do (see Birth 2008). What James did not fully articulate was that politicians and the state are bound to make plans to connect their present actions to future goals—such discourses are used to justify present policies. In contrast, calypsonians are not bound by such constraints and can prophesy and fantasize at will.

By the Carnival following the coup, interpretations began to take shape (Birth 1994, 2008). The most popular song of the Carnival, "Get Something and Wave" by Super Blue, mocked the government at the same time as the song created the character of a Spiritual Baptist woman, Mother Muriel, who had visions that Trinidad would "rise and reign." Another popular song, "Attack with Full Force" by Watchman, gave a detailed blow-by-blow comic description of the events of the coup mixed with social commentary. For instance, in one stanza he discussed how the members of Parliament were forced to drop their pants:

> I say if Bakr did shoot them down
> This year I'd of had nobody to pound.
> With all their big talk and arrogance,
> They belittle them by making them drop their pants.
> But I hear as soon as their trousers fall
> All the big boys in there turn out to be small.

Another song stands in complete contrast with Anderson's view of homogenous, empty time. Denyse Plummer's "La Trinity" links a "beautiful heritage" and a bright future determined by a divine power:

> No longer may I seem to be a paradise of glory
> Or a river of pleasure running out to the sea.
> But I take my refuge in a higher power, you see:
> My body will be restored and my heritage is still beautiful to me.
>
> Don't cry for me, my children, [Don't cry] for all the fire that falls upon me.
> Don't cry for me, my children, [Don't cry]. It was given to me to test me.
> But I will not be shaken, I am La Trinity,
> And there is no destruction, if you look you still see
> There is a mountaintop, there is the sea and sand,
> Child, I am your rock—I am your island.

The music and audience responses to the music of the Carnival that followed the attempted coup suggest that, despite the initial view of it as an anomalous event, Carnival placed it within a larger narrative of Trinidadian history in which Trinidad is peaceful with occasional outbursts of political violence. The unpredictable cycles of violence resemble the temporalities of permanence and unpredictable cycles documented in Trinidadians' representation of the Oil Boom. The narratives about the coup are easily identifiable and invoke past, present, and future through a teleological rubric of prefiguring—the future is brighter even if the timing of that future is unknown. While it indicates an empty near future, it is much like the young adults' narratives, where the near future was an extension of the present, whereas the rupture between present routines and ultimate goals was in the distant future. In contrast, IMF timelines and state plans follow a homogenous empty time in which the past, present, and future were separate and in which the power to do anything about conditions was denied.

The tragedy of the attempted coup was related to a triumphant narrative that linked with Trinidad and Tobago being a country where people are "loving," "free," and "cosmopolitan." The narrative of Trinidad and Tobago's eventual triumph developed during the same period in which the newspapers were speculating about the collapse of the ruling party, the NAR, and consequently the government. Those with whom I talked about politics described Abu Bakr as "stupid" and Prime Minister Robinson as "weak." The disconnect between how Trinidad's future was imagined and the leadership qualities of public figures was clear—political decisions did not prefigure the future and political leaders were disconnected from how Trinidadians imagined their nation. The qualities of Williams' inability to predict the future were associated with all political leadership. In effect, this temporality of unpredictable cycles combined with the judgment that politicians were incapable of anticipating these cycles undermined any political effort to persuade people that the government's plans and visions of the future were achievable.

In this light, contrary to how postcolonial temporalities are represented, it is not against linear time that postcoloniality pushes, but against the present prefiguring the future.

So the tension between narrative temporalities and homogeneous and empty time can be found in how the coup was discussed. The narratives about the coup are easily identifiable, and invoked past, present, and future through a teleological rubric of prefiguring and narrative logics, but when the topic turned to comments about economic and political change, the images of the past, the present, and the future emphasized a lack of agency about certain domains—the present economic condition was not prefigured by the past, and there was a sense of powerlessness to prefigure the future of the economy.

Temporal Counterpoints

There are also clear contrasts between the discussions of the economy generated by the IMF-imposed restructuring and the discussions of the future in relationship to the attempted coup d'état. The former emphasize uniform time and planning but evoke the narrative of political leaders' inability to predict the future. This counterpoint cultivates cynicism in any economic plan. In relationship to this narrative structure is always the idea of cycles of economic growth tied to unpredictable oil revenues. In these temporalities, there is always a connection between past events, the present, and the future, with the future being uncertain. In contrast, the narratives forged during Carnival in relation to the attempted coup place the coup outside of time. These narratives emphasize enduring qualities of Trinidadian society that the coup violated, and these narratives also emphasize that these enduring qualities will lead to a bright future.

To step back a moment and think about these narratives in relation to the future plans that emphasized sameness in the present followed by a rupture in the future, there is a parallel to what was discussed in the previous chapter. In discussing the past, "years aback" was a temporal frame that was always associated with a story with a moral message, whereas chronological time was associated with labor history and autobiography—making one's past intelligible to others. Likewise, the dominant songs of 1991's Carnival contain moral messages and hopes and indefinite temporalities of the future in contrast to the chronological and labor-oriented representations of the future found in the interviews. The temporal frames used to represent the past are duplicated in the temporal frames used to structure the future.

Empty homochronicity and the desire for narrative coexist and interact, and over time and context they shift in the extent to which they capture and organize consciousness. The choice between narrative and homochronism, then, is a choice between a self-representation of some connection between present action and future possibility, or an imposition of an emphasis on short-term questions because the future is beyond imagination.

So homogenous empty time provides an antinarrative structure that negates a sense of purpose not only within postcolonial societies, but also on a global scale. Its relation to ideas of agency is pernicious—it does not deny agency, but instead indicates that agency has little effect on the future. It becomes a way of imagining not only one's own future as beyond one's influence, but also the future of others. Maybe this contributes to the emphasis on death and nothing else in the elderly, and the young, newly unemployed man's desire to achieve fame through murder and self-destruction, or even the lack of will to address climate change, including the U.S. Department of Defense's explicitly adaptationist position (2014).

Consequently, the negation of purpose in homogenous time may be something that is most noticeable in how economies are imagined in postcolonial societies, but the ideological consequences of homogenous time must be recognized as global in scope. As I have argued in previous chapters, homochronicity is another feature of modernity that emerged out of the combination of science and globalization that is the heritage of the European Enlightenment. Its threat is that it cultivates a sense that actions do not matter.

THE POVERTY OF THE FUTURE

Why should homochronic temporalities be associated with cynicism about economic and political policies and, in contrast, prefigured temporalities with the imagination of the nation as a community? This does not fit Anderson's account. In his book *Imagined Communities*, Benedict Anderson places homogeneous empty time among the "cultural roots" of nationalism. Anderson's idea of temporality provides a phenomenological foundation for his overall theoretical framework. The shift in temporality he describes is a shift in "modes of apprehending the world" (2006, 22). In Trinidad, homogeneous empty time is not the sole temporality for apprehending the world, but only for specific domains of knowledge—domains that are associated with European ideas of human reason limited by chance, as opposed to ideas about destiny and fate. Is Anderson wrong, or is he a colonial occidentalist?

In this regard, what has happened to the game of whe whe in Trinidad is instructive. Until the attempted coup in 1990, each evening men gathered outside of a rum shop, in the community I studied, to record their bets, which were written on small scraps of paper. These scraps were then handed to a courier along with the money for the bets, and he would travel to where the local banker revealed the number between one and 36 that he had written down and then paid those who had chosen the winning number. Each number had a meaning. The banker ostensibly chose his number based on his dreams, and players would choose their numbers based on their own dreams, omens they saw, or their attempt to guess at the dream narratives of the banker. The man who first explained the game to me gave the following example: "You dream of a man being put in a hole, so you might play four for 'dead man' and 21 for 'mouth,' because a mouth come like a hole." Choosing a number could also involve group discussions. For instance, I witnessed the following exchange one evening: "I dreamed of your mother last night, [so] I playin' 'old lady' [number two]," to which the other man replied, "If you dream of my mother, you playin' 'queen' [number 24]!!!!" In another instance, a man saw an upside-down crapaud [a large tropical toad] in the road, and determined to play 31. His reasoning was that the number for crapaud is 13, but since the crapaud he saw was upside down, he should reverse the order of the digits—hence, playing 31 instead of 13. This man's brother saw the same upside-down crapaud but disagreed with the choice of 31—he related the crapaud to a bad dream about a tragic death that he had the night before, and based on the assumption that the crapaud was dead, he chose to play four, the number meaning "dead man." Often the number played the night before would be a factor in determining which number to choose. As another regular player explained to me: "City men don't know how to play whe whe like country people. In town, one banker play four [dead man] and then 18 [water boat]. If he do that in the country, he dead [his bank would be broken]. Everyone know that 'water boat' also means 'coffin,' and that after you dead, you need a coffin, so in the country everyone [those betting, not the banker] would play 18 after four, or maybe 21 [mouth], because a mouth come like a grave."

In subsequent visits, I found whe whe to have been taken over by the government as a form of lottery. With the government running the game, there are no longer daily, involved exegetical discussions of dreams and omens in relation to the game. It has moved from a game of divination about the dreams of a human banker to a game of chance—from a game about prefiguring to a homochronic lottery.

Critiques of Homogeneous Empty Time and the Imagination of History

Chakrabarty (1997) argues that homogeneous empty time treats time as belonging to nature itself rather than as the hegemonic construction it really is, and elsewhere is critical of histories that privilege a secular, European historiographic convention and theories over conventions and theories of non-Western histories when writing about the world outside of Europe (2000). These criticisms point to a conceptual danger of treating the uniform temporality of homogeneous empty time as natural. Indeed, the ontological foundation for homochronicity's claim to be natural is suspect. If, as Albert Einstein argues, time is relative to motion and space, then a time that is uniform and homogeneous across space is a fantasy, and simultaneity across space is, according to Einstein, an "illusion" (1992, 378). If this is the case, the signifiers of dates and clock times that ground homochronism are not only arbitrary, but refer to a signified that is also arbitrary; for instance, Greenwich, England, has no natural quality that makes it uniquely qualified to mark the prime meridian by which all clock time on Earth and missions into outer space are defined.

Homochronism is constitutive of claims of rational knowledge about nations vis-à-vis economic and political issues; it is the temporality that privileges the random over the destined. It privileges continuity and what Hodges calls *processual temporalities*—an emphasis on unity over time. Hodges argues that this temporality organizes scientific and ethnographic discourse, and denies the recognition of immanence (2014).

From Latour's perspective, homochronicity and its associated uniformity of time units is part of a broader tendency of being modern, namely, "to place the world of indisputable matters of fact outside history" (2004, 192). Latour contrasts this with the temporality of the nonmoderns, who embed knowledge in the anticipation of future possibilities (2004, 195–198). His solution to the dilemmas posed by the modern temporality that undergirds knowledge is a synthesis of the nonmodern experimentation in anticipation of the possibilities of the future while keeping track of the experimentation that has taken place (2004, 200).

Conclusion

Latour's synthesis does not work, however. It is an application of a timeless dialectical logic, and in applying timeless logical principles, it undoes itself. His critique demonstrates the limits of homochronicity, but does not eliminate this temporality. Even though homochronicity is limited, it still shapes thought and coexists and collides with other temporalities, such as those of narratives. Homochronism does violence to narratives. As Mulla has pointed out with regard to counseling rape victims in Baltimore emergency rooms, "The nurse, working by the clock and calendar, drains the vitality from the life narrative the victim works through in the aftermath of violence. Only when the clock, or the interview, stops, does time come to life as the victim is allowed to speak and tell the nurse and the advocate what matters to her" (2014, 73).

The ethnographic analysis of whe whe and Trinidadian responses to crises indicates an active imagination that deploys multiple narratives and that seeks a prefiguring of the present and the future. In contrast, the emergence of homochronism and its widespread acceptance is, in many respects, a product of the efforts of intellectual elites, and it is particularly potent in how economic theory and policy is imagined. From Newton's statement that environmental cycles are too irregular to measure time (1934, 7–8), and by implication discarding the rotation of Earth as a measure of the day, to the emergence of time zones that erase local time reckoning, to the adoption of universal time standards, and the use of clock time to commodify labor, the colonial project coincided with the emergence of a global hegemony of uniform, homogeneous time (see Landes 1983; Bartky 2007). This temporal uniformity is not merely a matter of Europe imposing something on the rest of the world, but as a brief look at the history of labor relations indicates, it is a matter of the imposition of temporal uniformity for purposes of control in any relationship of domination (see Roediger and Foner 1989; Smith 1997; Thompson 1967; Foucault 1977). Thus, the dissonance between homogeneous empty time and narrative time is not limited to former colonies but is global in scope, as is the lingering presence of each temporality when the other is emphasized. The manifestation of this tension, and the dissatisfaction of only recognizing one temporality when both are in play, can be witnessed in local Trinidadian interpretations of national crises, and even in a Trinidadian numbers game.

CHAPTER 6

Thinking Through Homochronic Hegemony Ethnographically

In one of the most quoted observations in the study of time, Augustine wrote, "What is time then? If nobody asks me, I know: but if I were desirous to explain it to one that should ask me, plainly I know not" (1997, Book 11, Chapter 14). Alluding to this statement, Paul Valéry observed, "Qu'est-ce que le Temps? —C'est un *mot*" [What is Time? It is a *word*] (1973, 1334, emphasis in original). The English word *time* is polysemic. It connotes moments, sequences, epochs, events, meter, rhythm, and simultaneity, pace, and the relationship of the present to the past and the future. As a result, when time is discussed, it is not always clear what is meant. For instance, there have been several works that deal with the temporality of the future (Guyer 2007; Lakoff 2008; Luhmann 1998; Miyazaki 2006; Robbins 2007), but for the most part they analyze the relationship of the future to the present, not the cultural ideas of sequence, epoch, and kairos that structure how the future is represented.

Failure to recognize the complicated relationship between the word *time* and what it signifies carries the risk of allowing tacit cultural ideas to become implicit ontological assumptions that affirm the normative, hegemonic order. As Sylvia Wynter observes, "The *normative categories* of any order … are normative precisely because the structure of their lived experience is isomorphic with the representation that the order gives itself of itself" (1984, 39). Even though anthropology is skilled at discussing diversity and differences, anthropologists are still culture-bound people who fall back on their cultural habits of communication. Consequently, the normative categories of time structure lived experience and representations of difference in anthropology. On the one hand, this can be critical for intelligibility to the discipline's academic audience, but on the other hand, it can distort and obscure the cultural differences that anthropologists seek to describe and understand. The habits of thinking about time are rarely interrogated in anthropology (Fabian 2002 is an exception), but lurk in the background to organize ethnographic representation. As a discipline well positioned to develop what Wynter describes as the "liminal categories" necessary to bring the artificiality of the normative categories into view (1984, 39), anthropology's accomplishments have been modest, at best, when it comes to addressing time. Instead, anthropological representation usually merely transforms myriad temporalities into normative post-Enlightenment European temporalities.

The temporalities associated with clocks and calendars are so naturalized as to be deployed without reflection even though they compel the adoption of a particular temporality that emphasizes uniformity and privileges duration. If one goes back to formative texts addressing the ideas of time now considered to be normative, one sees that these ideas of time are not viewed as natural, at all. Instead, they were matters of contestation. In effect, what is considered natural and normative is merely the ideas that won out in mostly forgotten disputes about time. In the thirteenth century, Roger Bacon complained that the Julian calendar "est intolerabilis omni sapienti et horribilis omni astronomo, et derisibilis ab omni computista" [is intolerable to all wisdom and horrible to all astronomy, and scornful toward calculation] (1859, 272). In the seventeenth century, in his pathbreaking book describing in detail the application of the pendulum to the measure of time, Christian Huygens wrote, "permanens magnitudinum mensura, quae nullis casibus obnoxia fit, nec temporum injuriis, aut longinquitate aboleri aut corrumpi possit, res est & utilissma, & a multis pridem quaesita" [a permanent magnitude of measure, which is not subject to chance nor injury to time from the destruction or corruption in length, is a thing most useful and sought for a long time] (1673, 151). Then, several decades later, Newton wrote, "For the natural days are truly unequal, though they are commonly considered as equal, and used for a measure of time; astronomers correct this inequality that they may measure the celestial motions by a more accurate time" (1934, 7–8). Bringing the possibility for such contestation back to the foreground is one element of combating the subtle privileging of the colonialist heritage of Enlightenment models of time.

As cognitive artifacts, calendars and clocks think for us (Birth 2012). Their cognitive function comes with the baggage of the assumptions in their algorithms. Most users are not aware of this baggage—their understanding is limited to what was learned early in grade school. That learning was focused on reading the representations, not on unpacking the logics that produced the representations. This temporal knowledge a young child obtains in the European-derived educational system reproduces the privileged position of the West.

Achieving Coevalness Through Communication

In *Time and the Other*, Johannes Fabian (2002) offers an antidote against this construction of privilege. His solution is the creation of coevalness in description. He focuses on the ethnographic interview, but again, his point is, in fact, more general. When representing the past, the present, or the future, a step toward over-privileging European temporalities is to connect one's own time to that of what is being described. That said, ethnographic interviews are a good context in which to witness the struggle for coevalness. Interviews serve as examples of the establishment of intersubjective time between researcher and research subjects. Intersubjective time has two meanings, however: shared experience in time, and shared temporal frameworks used to make communication intersubjectively significant. This point is demonstrated by two interviews in which I engaged during my first fieldwork—the topic of these interviews was the same, but my sense of being able to represent coevalness ethnographically with each person interviewed differed greatly.

The interview with "Raj" took place after over a year of field research. I had chosen him because he was engaging and articulate and willing to express his opinions and experiences. The interview was part of my field research about Trinidadian conceptions of time (Birth 1999).

KB: [T]he expression I want to start with is "Any time is Trinidad time."
Raj: "Any time is Trinidad time," ahm, that has only become a recent phenomena. Long ago, when they say you are going to take work at seven o'clock, you must be there at ten minutes to seven, and if you are not there you will lose a whole day's work.

After this, Raj told stories amounting to ten transcribed pages to bolster his point. A close reading of how Raj frames his response demonstrates the sophistication with which he handles temporality. His first response includes "that has only become a recent phenomena" and he immediately follows it with "long ago." Fourteen lines in the transcript later, he says, "I think the younger people are the ones who really exploit this 'Any time is Trinidad time.' The old people have grown up in a line of discipline." By this point in the interview, he has given me temporal frameworks defined in terms of the past, the present, and generations. He uses this temporal matrix to elaborate on differences in discipline between young people and old people. To emphasize his point, he introduces a story about how the old people were raised with the temporal frame: "Now, the discipline long time…" Each of his stories is temporally framed along two dimensions: the present versus the past, and the people who are presently old versus the people who are presently young. For instance, in the following example, he tells a story about a recent meeting, but uses it to build the contrast between young people and old people:

> We organize a woman's group. When it was, we organizing the meeting for five o'clock, by six o'clock being home, you must be home by half past six because husband always suspicious of their wives, even go to the meeting and so on. When it quarter past five, many of them walked to the meeting. When it quarter past five some started to show. One or two came in on time—the meeting was to be at five. Some came after five. Some came half past five. When it was half past five and we had enough people, I said, "Ladies, I'm sorry some of you must have come from very far, but this is a woman's affair, and let me tell you as a man how your husband will think. We start a meeting half past five and end at seven, he will say, he will not want to understand that the meeting started late, so we will not keep any meeting today, but next week Friday, same day, at five o'clock," and everybody there quarter to five ready for meeting. Those are the woman and the older people. The younger people, I don't think that would have made an impact. You still find some, 25 percent, coming in after meeting started.

At another point in the interview, he was lamenting the growing power of unions and their effect of eroding punctuality and framing it in terms of "now" versus "long time," a Trinidadian idiom used to refer to the distant past: "Now, the discipline long time, where the chief overseer on the road could send you back home and you have to lose a day work—if he do such a thing now he will be answerable to the union." His multidimensional temporality of generations and "now" versus "long time" allowed him to compare the behavior of the old people when they were young to the behavior of contemporary young people. Indeed, he was evoking the morally tinged "years aback" temporal frame discussed in Chapter 4. With this temporal frame, Raj could offer his moral opinions on a range of topics from school, to cricket tests, to work, to local meetings. Every setting is oriented by his matrix of present/long time and old people/young people.

The coevalness I felt was not my creation—it was not the result of my being able to place his experience into a temporal framework that I crafted. Instead, the coevalness I felt resulted from Raj's ability to convey his temporal frameworks to me. In Fabian's work, the field-worker's sense of coevalness was not addressed. He emphasized the issue of time as an epistemological contradiction between allochronic discourses and empirical research that unfolds through sharing time in communication with research subjects. The interview with Raj was meaningful not simply because we shared a moment in time, but because of how he related our shared moment in time to his life and his sense of his past. Raj's interview suggests that the issue of coevalness is not only epistemological but also phenomenological.

The extent to which the coevalness I felt was Raj's doing, and not mine, is demonstrated by an interview in which I felt allochronic—outside of the time of the person to whom I was talking. "Robert" was difficult to interview. Each question I posed was greeted with a short, vague response. I had wanted to interview him because of his age and work history, and because in other social contexts he was talkative. We had known one another for a year and a half, and had participated in conversations about many topics. I was following the same protocol I had with Raj.

K: I have heard this expression: "Any time ..."
R: "... Is Trinidad time."
K: What does it mean?
R: They does have a way they does say like how you know when you tell somebody come and check you say, "Any time." They does say, "Any time is Trinidad time." That is how most of the people really does use it for. Like, "Any time is Trinidad time."

K: What do people mean when they say "jus' now"?
R: Well, sometime, somebody might give you something and they mightn't right away be able to come to attend to somebody, so they tell you, "I comin' jus' now ..."

K: What do people mean when they say "long time"?
C: Like long time, like fifteen years aback and twenty years aback and thing. Everybody say "long time ..."

Robert provided no temporal frame, and since I was not fully cognizant of this being what I was missing, I was not equipped to get him to develop his thoughts in a way that I could grasp. Even when he discussed the idiom "long time," which refers to the past, Robert did not provide a temporal frame that I adequately understood—he used the phrase "years aback." In Chapter 4, where I discuss how Trinidadians temporally structure their discussions of the past, I argue that, of the four commonly employed means of temporal structure—dates, stage of life, significant historical events, and "years aback"—"years aback" is unusual. The other three temporal structures are often associated, thereby creating dense networks of temporal orientations for any story. In a close reading of the interview transcripts, "years aback" rarely occurred with the other temporal structures. Instead, it tended to be used to elucidate general principles, usually moral points, that were true in the past and remain true in the present. Narratives framed in terms of "years aback" have qualities similar to how Fabian describes allochronism—a time outside the normal flow of time. Robert's use of this phrase to frame his discussions without including any didactic narrative left me temporally disoriented. This was compounded by his use of "you know," "something," and "everybody"—indicating that he used a restricted code—so that much of what he could have said, he did not (Bernstein 1971)—and by the fact that I did not have the knowledge necessary to fully engage him in the interview. Robert and I were in the same place at the same time, but we were not coeval in the same way Raj and I were coeval.

Fabian suggests that "for human communication to occur, coevalness has to be *created*" (2002, 30–31, emphasis in original), but my experiences with Robert and Raj, and my ethnographic work on how Trinidadians represent their lives in time (Birth 2006), have led me to be believe that communication is not a sufficient condition for coevalness to emerge. The differences between my interviews with Raj and Robert complicated for me what the creation of coevalness might mean, and suggested to me that a shared duration of communication was not the same as a sense of a shared phenomenological sense of time. These interviews also caused me to think that a subtle dimension of rapport is the mutual intelligibility of the temporal organization of narratives—in effect, that to create coevalness and to incorporate it into ethnographic representation required more than a demonstration that I was in the same place at the same time as those with whom I worked.

The problem Fabian raises is far greater than can be solved by establishing coevalness. Coevalness is still bound by the very concepts of time that distort research. Still, thinking through the problem of temporality by thinking about ethnographic representation is useful simply because ethnography struggles with the representation of alterity. Writing ethnography always involves an effort to craft a representation that portrays difference in terms of sufficient familiarity that the reader can understand. It is a multistage process that includes the interview encounter, but also the transformation of that encounter into field notes (see Sanjek 1990), and then the strategic culling from those notes in the crafting of ethnographic description.

Beginning with the interview, the problem of temporality becomes apparent. The interviews I have described suggest that coevalness is not a creation of the ethnographer, but a creation out of the intersubjective relationship of the ethnographic encounter. Coevalness has epistemological, rhetorical, and phenomenological dimensions. Yet, even if coevalness unfolds in a field situation, its translation into an ethnographic account still faces four major challenges: the split temporalities of the ethnographer; the multiple temporalities of different histories; the culturally influenced, if not constructed, phenomenological present; and the complicated relationship between culturally variable concepts of being and becoming and cultural concepts of time. Consciousness of these challenges raises awareness of the risk of homochronism—a displacement of those people who are ethnographically represented out of their temporality and their assimilation into academic discourses of history. Homochronism can be subjected to criticisms that parallel those Fabian made of allochronism—both homochronism and allochronism are tropes that distance the Other through placement into post-Enlightenment temporal constructions.

The First Challenge of Coevalness: Ethnographers' Split Temporalities

Whereas Fabian emphasizes how the ethnographic present distances the ethnographer from the subject of ethnography, he does not explore the complicated relationship of the ethnographic present to the temporality of the ethnographer. In discussing chronotopes, the narrative relationship of place and time, Bakhtin explores the relationship of the author to the narrative and points out that "[t]he represented world, however realistic and truthful, can never be chronotopically identical with the real world it represents, where the author and creator of the literary work is to be found" (1981, 256). The author's presence in ethnographic representations is often a split image: the author as writer at the time of writing, and the textually represented field-worker whose temporal and spatial relationship to the ethnography is complicated. With regard to texts in which the author is represented within the narrative, Bakhtin states, "He can represent the temporal-spatial world and its events only *as if* he had seen and observed them himself, only *as if* he were an omnipresent witness to them. Even had he created an autobiography or a confession of the most astonishing truthfulness, all the same he, as its creator, remains outside the world he has represented in his work" (1981, 256, emphasis in original). This point is illustrated by Crapanzano's discussion of Geertz's opening narrative in his essay "Deep Play" (Geertz 1973; Crapanzano 1986). Crapanzano focuses on Geertz's story of escaping a police raid on a cockfight as a rhetorical means of establishing Geertz's ethnographic authority, and in doing this, Crapanzano highlights the difference between the anthropologist as a character in an ethnographic representation and the anthropologist as the producer of that representation. In "Deep Play," then, Geertz is both a character in a story and the author of that story, but Geertz-the-field-worker in the story is portrayed with uncertainty and anxiety about the future that Geertz-the-author lacks. From a textual perspective, one could ask whether Geertz established coevalness with himself—a point made by Rabinow, who wrote, "We remain out of time not only in the sense of a refusal of coevalness with the Other in Johannes Fabian's terms, but in a strict sense, in a refusal of coevalness with ourselves" (1985, 360). So, whereas recent anthropological rhetoric eschews an ethnographic present in favor of representations of the ethnographic presence of the author as field-worker, the implicit split image of the two temporal existences of the ethnographer as narrator and character in a narrative remains a challenge for representations of coevalness. These conventions rarely relate the author to history, or even to the author's pre-fieldwork past, but suspend the ethnographer in time, in an effort to establish coevalness between the ethnographer and the Other.

The Second Challenge of Coevalness: Multiple Temporalities of Multiple Histories

Fabian's *Time and the Other* is often used to support the incorporation of history into anthropological representation. Behar credits Fabian's ideas with "the infusion of a deeper level of historicity into ethnography" (1986, 6). For Borneman, Fabian's *Time and the Other* encourages the following perspective: "By striving toward a fuller historical consciousness, we anthropologists are doing much more than merely acknowledging the historicity of our productions ... Our understanding of historical space is more than a prelude to knowledge of the objective world; it is also a form of knowledge about events of cultural significance necessary for a self-articulation and definition in the present" (1998, 146–147). The significance of Fabian's ideas has been extended beyond cultural anthropology to include archaeology. Hodder writes: "Involving ethnographers will hopefully assist archaeologists to shy away from assuming an equation between 'local' and fixed or indigenous. A reflexive approach to the local involves seeing how it is historically constructed. The local may not be an 'authentic' voice that can be used uncritically to make sense of the past in that locality" (2003, 63). In relation to the intersections of ethnography, ethnohistory, and archaeology in the study of Mesoamerica's past, Chance uses Fabian to articulate a concern about an "ethnographic past" that also distances "the Other" from the anthropologist: "By giving short shrift to history, we may attribute a greater antiquity to beliefs and practices than they in fact possess" (Chance 1996, 392).

The invocation of historicity and the use of historical narratives is seen as a means of avoiding allochronism. Yet, many ethnographies equate history, a representation of the past, with historicity, a representation of a connection to the past, and contextualize their representations using Western historiographic ideas. In this case, history is used as the ground for shared pasts, but in personal encounters, it is not such general history that is relevant, but locally embedded history. The distinction Glissant makes between *H*istory and *h*istory (1989) emphasizes the difference between the assertion of transcendence of History as represented in Western discourse and the experience of fragmented pasts that is found in local history in the Caribbean. Glissant's distinction resonates with Foucault's contrast between "traditional history" and "effective history." For Foucault, traditional history "aims at dissolving the singular event into an ideal continuity" (1984, 88), and effective history "deals with events in terms of their most unique characteristics, their most acute realizations" (1984, 88). These are not only different views of history, but they suggest how techniques of creating history organize how the past is thought. History dominated by linear chronology homogenizes time and essentializes the unique qualities of events. It places events within a general continuity of uniform units of time, as opposed to leaving them within the sequences and rhythms from which they emerged.

If identities emerge, in part, out of the relationship between local discursive practices and local histories, then does linking them to "History," in Glissant's sense, or to traditional history, in Foucault's sense, establish coevalness? Moreover, the contents of History are themselves contested (Chakrabarty 1992, 1997, 2000; Trouillot 1995). Consequently, history cannot easily be assumed to be the past of intersubjective significance for communicative encounters between a field-worker and informants.

When I interviewed Robert, I encountered this problem. The local history of which Robert was conscious is not the history represented in works on Trinidad's past. The representations of Trinidad's history that shaped my awareness of Trinidad's past emphasized the sugar plantation economy and ethnic politics. Robert lives where cocoa is the main crop and the effects of ethnic politics refract through locally based agricultural and labor interests (see Birth 2006). Knowledge of Trinidad's past as recorded in academic discourse is not identical to Robert's knowledge of his past as part of these historical currents. Thus, even when one refers to history, this is not sufficient to establish a common present in the ethnographic encounter, because the ideas of the past that the ethnographer and the informant bring to the situation are different.

The distinct temporalities of different histories and historicities compound this problem. All representations of the past involve temporality. Raj was successful in creating coevalness because of the sophisticated matrix of temporal frameworks he used to contextualize his stories about the present and the past. Subsuming ethnographic representation into any particular history avoids the problem of allochronism, but creates homochronism—a single all-encompassing set of temporal tropes. History (in Glissant's sense) relies on a post-Enlightenment epistemology of time—the Christian chronology, an emphasis on change, and a distinctive periodization often subtly defined and punctuated by European and North American conflicts. These temporal tropes organize and orient events in time, and they are part of what Fabian describes as "attempts to secularize Judeo-Christian time by generalizing and universalizing it"

(2002, 2). Even if one attempts to write effective history, in Foucault's sense of the concept, employment of these temporal frames places such effective history into the sort of ideal continuity Foucault associates with traditional history.

By the time of the publication of *Time and the Other*, many ethnographies already included history, so Fabian's book reinforced this trend. For example, in *The Devil and Commodity Fetishism*, Taussig (1980) creates this sort of coevalness in order to link folk religion to Marxist ideas of class consciousness. In discussing the plantations, he offers a narrative punctuated by dates and periodizations to document the relationship between Europeans, Africans, and Indians:

> The Inquisition was founded in Cartageña in the early seventeenth century. (1980, 42)
>
> ----------
>
> Writing in 1662, the chief inquisitor attributed much of the sorcery and idolatry in the mining districts to the heedless materialism of the mine owners. (1980, 43)
>
> ----------
>
> In 1771 the Bishop of Popayán, capital of the Cauca region of southwest Colombia, complained bitterly that his attempts to catechize the slave and prevent their being worked on Sundays and feast days encountered the firm opposition of the slave owners. (1980, 45)
>
> ----------
>
> In the opinion of Ramón Mercado, native of Cali and Liberal party governor of the Cauca region between 1850 and 1852... (1980, 45)

This demonstrates the frequency with which historical chronology and periodization is used to frame representation. Taussig's multilayered temporalization is not limited to years—the narrative is further structured by "periods" in which local trends are linked to supralocal history: for example, "the Inquisition."

In contrast, Rappaport's discussion of the construction of history and time among the Páez of Colombia (1998) suggests the extent to which Taussig might be imposing his temporal and chronological sensibilities on his ethnographic subjects. Rappaport demonstrates that the histories of the Páez do not follow the same narrative and chronological form of Western, academic history that emphasizes linear periodization and causal chains. Instead, Paéz historicity includes a variety of genres with different narrative conventions, some of which emphasize the motifs that make sense due to patterns of repetition rather than chronological proximity. How does one think through the challenge of multiple histories in connection to coevalness? Does homochronism artificially obscure important differences? In the previous chapters I have argued that it does.

The Third Challenge of Coevalness: The Phenomenological Present

The idea of the present invokes ideas of the past and the future. The ethnographic encounter is one of the intersections of different phenomenological, subjective pasts in an intersubjective present. Discussions of the phenomenological present affirm that it is closely tied to memory and imaginations of the future. Different pasts and futures lead to different experiences and constructions of the present (James 1996; Mead 1932). Even when invoking an idea of collective memory, Robert and I had only slightly overlapping memories, and consequently, our shared present was an intersection of different temporal subjectivities, not coevalness. As Fabian acknowledges, "*Somehow we must be able to share each other's past in order to be knowingly in each other's present*" (2002, 92, emphasis in original). This poses a challenge to the emphasis on phenomenology and cognitive universals—these are insufficient to create a shared past.

Cole's work on memories of colonialism in Madagascar provides an example of the complex interaction of context-dependent remembering with an ethnographic portrayal of the interview encounter. In *Forget Colonialism*, she reports going to the field with the intention of studying "how historical events were experienced in terms of everyday consciousness" (2001, 2). In particular, she was interested in the 1947 rebellion against French colonial authority. She recounts that when she arrived at her field site, the presence of the violent colonial past seemed absent, but that instead, the relevant past was dominated by "the moral economy of cattle sacrifice" and concern about ancestors (2001, 3). She admits, "I might have concluded that the importance attributed to the colonial period by Western anthropologists and postcolonial theorists had more to do with their own preoccupations than those of the formerly colonized" (2001, 5). Later into her fieldwork, Malagasy elections prompted open and vivid discussions of the colonial past and the violence of 1947 (2001, 2006). In effect, the relationship of the past to the present was shaped by the concerns of the present, and her initial position of thinking in terms of the violent colonial past did not fit with the present she encountered of the cycle of cattle sacrifice.

Koselleck (1985) argues that the ways in which the future is imagined and linked to the present vary historically. By implication, futures vary culturally, as well. Bourdieu's discussion of Kabyle concepts of the future corroborates Koselleck's point: "[N]othing is more foreign to the pre-capitalist economy than representation of the future (*le futur*) as a field of possibles to be explored and mastered by calculation" (Bourdieu 1979, 8). Bourdieu distinguishes between *le futur* of capitalism, and the forthcoming future (*un à venire*) grounded in the present, and argues that the latter is typical of precapitalist societies (1963, 61–62; 1979, 9). Bourdieu's claim to be studying "traditional" Kabyle has been challenged by Goodman (2003), who points out that he worked with refugees during the Algerian Revolution, but even if Bourdieu's portrayal does not reflect traditional Kabyle concepts of the future, it still suggests cultural variability in concepts of the future. There is additional support for such claims in the vast literature on achievement motivation and future orientation in psychology (see Ashkanasy et al. 2004 for a review). This literature uses psychological tests cross-culturally to discern differences in how people think about the future. Often, these studies have practical intentions of explaining different levels of economic success in capitalist economies. The methodological problem with the studies is the extent to which they implicitly adopt ideologies of time and planning associated with European and North American capitalist values, and it is unclear whether the tests measure how people conceptualize the future or the extent of their differences from European and North American representations of it. Despite this significant flaw, this literature does corroborate Koselleck's claim of historically and culturally contingent relationships with the future. But it only does so with a limited imagination of the temporalities that structure the future—indeed with the implicit assumption that the future is structured by homogeneous, uniform, linear time. The last chapter demonstrated how impoverished a view this is—multiple temporalities structuring multiple futures coexist.

Ethnographic coevalness, then, rests on an assumption of not only an intersubjectively shared present, but, by implication, an intersubjectively shared past and future, as well. This does not mean sharing identical pasts and futures, but sharing sufficient common knowledge about the past and the future to make communication in the present intelligible (Birth 2006). Sharing in a communicative event does not automatically generate the sort of intersubjectivity necessary to establish coevalness. Robert and I did not have this shared past or future, and as long as he spoke in a highly restricted code, we could not have an intersubjectively shared past sufficient for me to make sense of what he said. In contrast, Raj made a great effort to make his past intersubjectively shared between us.

Culturally divergent memories and anticipations of the future raise another question of phenomenology in relationship to conceptualizing coevalness: To what extent is the phenomenological present shared between readers, writers, and ethnographic subjects, and consequently is the "present" of a communicative encounter a sufficient means for establishing coevalness?

Schutz's contrast between contemporaries and consociates describes another dimension of the concept of the phenomenological present. Consociates share a directly experienced social reality, whereas contemporaries could share this reality if they chose to do so, but do not (Schutz 1967, 142). According to Schutz, contemporaries understand one another through "anonymous processes" that involve the understanding of "social reality in general" (1967, 183). The use of history in ethnographic research generally employs accounts that are not specific to the individuals studied, the ethnographer, or the ethnographic encounter, but relate to a larger social context that is more general and envelops ethnographic moments—a translation of the past of the individuals into a past of their "social reality in general." An anonymous and general past can be applied to all ethnographic subjects, but that hardly seems to be the coevalness Fabian seeks, and it is not the coevalness Raj created. Yet, a consistent challenge in ethnography is that published histories are rarely straightforward records of the past of the location of fieldwork (Trouillot 1995). Often, their use involves the application of national or regional studies to a specific locale— an application of a "social reality in general" to contextualize "directly experienced social reality." When using a general past of a nation or region in place of the unique past of a location of research, does ethnography transform consociates into contemporaries in order to create coevalness? The common rhetorical techniques used to create coevalness suggest that this is the case. By placing both the Other and the ethnographer in the history derived from scholarly studies of the past, coevalness seems possible, but such history is distanced from both the lived experience of the ethnographer and the ethnographic subjects. If coevalness is meant to overcome the creation of distance between anthropologists and those they study, then the use of homochronic history is just as inadequate, albeit in different ways, as the use of an allochronic ethnographic present.

THE FOURTH CHALLENGE OF COEVALNESS: TIME AND ONTOLOGY

In her extensive review of the anthropological literature on time, Munn writes: "When time is a focus, it may be subject to oversimplified, single-stranded descriptions or typifications, rather than to a theoretical examination of basic sociocultural processes through which temporality is constructed" (Munn 1992, 93). Munn's statement seems odd. After all, anthropologists have written extensively on time, and this literature is complemented by work in other disciplines. Yet, a close examination of the literature on time in the social sciences and humanities supports Munn's claim, and reveals that the literature on time tends to develop only one of the following three themes at a time: a demonstration of the cultural variability of concepts of time; a concern with ontology and time; or a concern with how time organizes thought.

Much classic work in anthropology, such as Hallowell (1955), Geertz (1973), and Evans-Pritchard (1940), has emphasized cultural variability in concepts of time. In the cases of Geertz's "Person, Time and Conduct in Bali" and Evan-Pritchard's *The Nuer*, the cultural variability of time played a central role in representing the alterity of those studied.

Such an emphasis on cultural differences conflicts with those who seek to emphasize transcendental elements in the human experience of time. Concern with transcendental features of time in opposition to cultural variability falls within the tradition of European philosophy. For example, Heidegger (1962, 1989, 1992) develops the concept of "*das Man*." This concept often has been translated into English as "they," but it also means "one," "everyone," and "we" (Heidegger 1992, 327). In effect, it suggests a density of the relationship of intersubjectivity and subjectivity that no single English word adequately captures. *Das Man* can imply a non-anaphoric, imagined "they," or the collective. Heidegger argues that the effect of *das Man* on being-in-the-world is to create an alienation that "*closes off* from Dasein its authenticity and possibility" (1962, 222, emphasis in original). By pitting culture against authenticity, Heidegger poses a significant conceptual challenge to an anthropological perspective that links culture to authenticity.

A compromise perspective seeks to balance universal qualities and cultural variations in thought. This approach is derived from Kant's idea of time as an analytic judgment that plays a crucial role in forming synthetic judgments (2003). It has taken its most recent forms in works on how time structures narrative and representations (Ricoeur 1984, 1985, 1988; Carr 1986; Bakhtin 1981). Ricoeur develops the idea of how concepts of time structure authoring, the text, and the interpretation of the text. Like Kant, Bakhtin (1981) also makes time a fundamental dimension of humans' abilities to tell and to understand stories. Whereas Ricoeur and Bakhtin suggest that the use of time to structure stories is a human universal, unlike Kant they do not assume that the concepts of time used are Newtonian, sequential concepts. Instead, both Ricoeur and Bakhtin suggest variability in concepts of time even as they stress the necessity of concepts of time in making narratives meaningful.

Balancing cognitive temporal universals with cultural temporal variation seems to be the most amenable approach for anthropology. By avoiding the problem of ontology—particularly Heidegger's formulation of humans' concepts of being having a temporal dimension—one is able to avoid a conflict between a transcendental ontology of time and cultural variation in concepts of time. Yet, many societies have developed ontologies of time, and to refer to these ontologies as "cultural concepts" becomes a way of avoiding the task of relating them to European philosophical treatments of time, much less using them to challenge European ontological ideas about time.

Some ethnographic work has engaged the relationship of ontological ideas of time and cultural variation in concepts of time. In her work on Oman, Mandana Limbert confronts the connection between Omani ideas of time and being-in-the-world. Omanis have an anxiety about the future that disrupts their sense of the present. Their sense of the present is viewed as miraculous, and thereby a disruption, rather than a repetition of the past (Limbert 2010). Their sense of who they are is grounded not only in this miraculous, anomalous present, but also in their connections to the past and the future. Yet, Omanis do not comfortably fit within European imaginations of ontology. The disruption in time is not the termination of their being in the future, but the manner in which the present is temporally anomalous—the present is viewed as a termination of their being in the past, and when the present ends, they may revert to what they were. This is not a cyclic concept of time, but one of enduring Omani characteristics in relationship to which present conditions are viewed as unusual.

Another example of the difficulty of relating cultural variability to a transcendental concept of being-in-the-world comes from Australian Aboriginal concepts of "The Dreaming" or "Dream-time." McIntosh refers to the link between this concept and the idea of being able "'to see the eternal'" (2000, 38). Hume represents this concept as "the sacred knowledge, wisdom and moral truth permeating the entire *beingness* of Aboriginal life" (2004, 237). Stanner describes The Dreaming as existing outside of time: "[I]t was, and is, everywhen" (1979, 24). In Aborigine ontology, being involves relations between this concept of the eternal and the waking experience of cycles and change.

The challenge of Omani concepts of time and Dream-time is that these concepts shape the relationship of being-in-the-world and identity to the past and the present. As Davis points out, concepts of time "are a raw material for the production of thought about the past" and, since concepts of time differ, so do the pasts that these concepts shape (1991, 4). Consequently, the consciousness of time is central to being for all humans, and this suggests that consciousness of culturally shaped times creates different senses of being-in-the-world. The problem, then, for anyone engaged in work that confronts cultural concepts of time related to ontological ideas of being-in-the-world—the challenge of Omani and Aborigine ontologies—is that the scholarship of cultural variability in temporal ideas can only awkwardly be coupled with the scholarship on ontology. European philosophy's emphasis on the transcendental component of the relationship of time to being-in-the-world collides with an emphasis on cultural variability. Such a tendency to eschew the cultural to discover a transcendental ontology creates complications for applying the third tradition concerning ways of knowing to understanding the relationship of time and ontology cross-culturally. Time, then, is a topic that raises all the issues associated with cultural relativism, but does so in a fashion that is also related to the problem of being.

Anthropology's Love of History and Fear of Time

The challenge to ethnographically understanding time and being suggests that anthropology must confront two related latent tendencies in its treatment of time. First, much contemporary anthropology desires to document and see change, an attitude that Fabian suggests is shared by all "Western social science" (2002, 145), and the anthropological study of time emphasizes temporal variability, and consequently, alterity. The difficulty of reconciling these two positions of a pancultural emphasis on change and a culturally specific emphasis on alterity has emerged in several important works on the anthropology of time. Bloch (1977) reconciles them by distinguishing ritual time, which is culturally variable, from sequential practical time, which is pancultural. Gell (1992) adopts a similar perspective by suggesting that a sense of time based on preceding and forthcoming (known as A-series time) is fundamental to experience, but becomes culturally elaborated in many different ways.

Munn (1992) is explicitly critical of Bloch's privileging of practical time over ritual time because it emphasizes an "'empirically' derived cognition that has somehow bypassed any sociocultural construction of reality" (1992, 100). Kermode, in his discussion of the importance of a concept of the ending in narratives, is even more condemning of allowing time to be conceived only in terms of succession: "To see everything as out of mere succession is to behave like a man drugged or insane" (2000, 57). His argument is that while time may involve succession, narratives about what occurs in time involve senses of beginnings, middles, and ends. Moreover, the middle, when the action takes place, is laden with the issue of the fulfillment of time as the plot moves toward the ending.

If coevalness requires not merely a shared duration of time, but intersubjectively shared concepts of time and associated ideas of beginnings, middles, and endings, then according to Munn and Kermode, and contrary to Bloch and Gell, it is not possible to rely on a bare-bones concept of succession as the ground for meaningful intersubjectivity or being-in-the-world.

An illusory way out is to ethnographically represent cultural variability while using a temporal structure of historical change that fits academic conventions to frame the ethnography. Such illusory coevalness not only assumes a shared past, but also claims a shared sense of being-in-the-world grounded in academic history—a very culturally, contextually, and temporally specific mode of thinking about the past. It forces both the ethnographer and the Other into academic representations of the past, being, and identity. Much of the use of history in anthropology does this. Rather than confronting the issue of time, the relationship of the past and the present is assumed, and European dating systems provide the orientational frameworks for structuring the past. Through a web of dates, global events, and local histories, the past is used as the context for understanding the present, with attention paid to how cultural representations of the past are structured by present concerns. To say that Plato died in 347 BC invokes the dating system devised by Dionysius Exiguus in the sixth century AD and the pre-AD dating conventions developed by Bede. We place all events and people prior to Dionysius and Bede into our conceptions of time and chronology, not theirs—Plato's death is represented by a date defined in relationship to an event (the birth of Jesus) that Plato could not have foreseen. Ethnographically, it is easy to impose such temporal frameworks without reflection.

Also, for all the extolling of cultural variability, when it comes to the fundamental issues of time and ontology, cultural anthropology prefers to make change, not conservatism, transcendent and absolute. This seems consistent with the postmodern condition that emphasizes transience, fragmentation, and instability (Lyotard 1984; Jameson 1991). Temporal conceptions that emphasize enduring characteristics, everlastingness, eternity, or infinite repetition are disconcerting precisely because they disrupt a sense of change as central to being-in-the-world.

In this regard, it is important to note that Fabian's criticism of allochronism was that it was in the service of Western social scientific theories of change. Placing of the Other outside of time occurred because "theoreticians and apologists of a new international order perceived the need to safeguard the position of the West. The necessity arose to provide an objective, transcultural temporal medium for theories of *change* that were to dominate Western Social Science in the decades that followed" (2002, 144–145). Whereas allochronism was often in the service of teleological concepts of change, homochronism need not be teleological. To exchange teleological allochronism for homochronism, but still to privilege theories of change that "safeguard the position of the West" does not address Fabian's concern. The problem is that for many people in many societies, change and stasis stand in relationship to one another.

Recognition of the ways in which change and stasis seem mutually disruptive is a widespread aporia. This problem is central to Augustine's struggle with the concept of time in his *Confessions* (1997). For him, God was eternal and unchanging and the challenges of temporality were due to the changing condition of humans. For Augustine, the understanding of change only unfolds in contrast to eternity—or, as Ricoeur describes it, Augustine's aporia is the relationship of *intentio* to *distentio animi* (1984). Augustine uses this to contrast God, whom he links to *intentio*, with human life—a process of maturation and transience characterized by *distentio animi*. In addition, Augustine was intent on demonstrating the superiority of his theory of time that combines linear time with a concept of eternity over representations of time as cyclic (1997,

Book 12, Chaps. 13, 14, and 17) and a concept of infinite linear time without beginning or end (1997, Book 4, Chapter 4), a temporal theory that can be attributed to Aristotle (1936), although since Augustine had no Latin translations of Aristotle, he was probably referring to Epicurean philosophy or Manichean discussions of this idea of time. Hegel's philosophy contains similar issues, with his emphasis on change in relationship to the Unchangeable and Absolute Idea (1977). More recently, Deleuze has wrestled with similar problems in his study of the relationship of difference and repetition (1994). This is not simply a problem within the intellectual tradition that stretches from Plato to European philosophy, but it is a problem within traditions in other parts of the world, as well. Peel's description of Ijesha concepts of history (1984) juxtaposes a logic of time-transcendent stasis in *itan*—a form of social charter—with a logic of succession and change in *qwa*—the representation of the succession of rules. Balinese concepts of time also address the aporia of relating change and repetition under the single concept of time (Geertz 1973; Howe 1981), in a way that seemingly privileges repetition over change. Miller, in his work on Trinidad, also notes a tension between what he calls transience and transcendence (1994). My own engagement with Miller's terminology initially focused on his link between these concepts and ethnic stereotypes in Trinidad (Birth 1999, 16–17), but the global distribution of attempts to relate a sense of change with a sense of eternity/stasis/repetition suggests that what Miller documented in Trinidad, which I also noted, might have been a complicated grafting of this aporia into constructions of ethnic differences. In any case, the transcultural, if not pancultural, distribution of such aporias, along with the great cultural variability in the solution of such aporias, makes it difficult to imagine an anthropological theory of time or representations of time with universal applicability. Likewise, there is a need to contextualize coevalness in very local, culturally sensitive terms, rather than through the imposition of the temporal concepts that are often hidden, yet important, assumptions in anthropological and historical scholarship.

Conclusion and Alternatives

The homochronism commonly invoked to create a sense of coevalness is consequently an imposition of one historically and culturally contingent, and presently powerful, temporality on both the ethnographer and the subjects of ethnography. Whereas much of this discussion has focused on ethnography, the discussions of Derrida, chronobiology, and labor economics suggest that time poses a multidisciplinary problem. In ethnographic work, ethnographic representations of temporality create four challenges: the split between the ethnographer's experience of time and the tropes for representing the ethnographer; the existence of multiple histories; the diversity of ways in which the relationship of the past and future shapes the phenomenological present; and the diversity of ethno-ontologies. These challenges can be translated into other fields, whether it is the scholar's control over the texts that are analyzed, or the scientist's control over variables in the laboratory setting. Since the construction of temporalities is often outside of the consciousness, its influence on research goes unnoticed—time is experienced as natural and beyond question.

Current homochronism hides these problems by creating a sense of a shared, general history. It is a history that makes everyone into contemporaries, in Schutz's sense, and a history that privileges change. Fabian responds to his worry about "a program for the ultimate temporal absorption of the Other," what I have called homochronism, with a desire to see "what can be said, positively, about coevalness" (2002, 154). In retrospect, his critique did not prevent homochronic writing. Since the problem is with how time is represented, ignoring the choices involved in representing multiple, coexisting temporalities does not solve the problem.

That said, there has been some work that provides alternative conceptions of time for understanding non-European power relations, and these have created interesting ways to represent coevalness. The already-mentioned work of Rappaport to capture Páez' sense of their past is one example (1998). Sahlins' study of the death of Captain Cook is another (1985). Crucial to Sahlins' argument is the claim that Hawaiian cosmology and politics were linked in such a way as to promote Cook to a position of honor during one part of the year, but to view him as a threat to order during a different part of the year. Obeyesekere (1992) has criticized Sahlins' argument on the grounds that Hawaiians were reacting to an overbearing British explorer, and that an appeal to cosmology provides a weak, if not a ridiculous, explanation for Cook's death. If the Hawaiians stood alone in the apparent link between concepts of time and political structures, Obeyesekere's argument would be powerful, but the Hawaiians are not alone. In a study of Mayan political organization, Rice argues that the connections between cosmology, calendrical reckoning, and power were crucial features of the Maya (2004). The early medieval debates about chronology and calendars also closely linked secular power, cosmology, and the control and awareness of time (Borst 1993). It seems post-Enlightenment rationalism severs time from eternity and precludes cultural ontologies that do not do the same. Works that confront this problem explicitly represent temporalities that contrast with post-Enlightenment rationalistic times. These alternative temporalities often address either cultural ideas of eternity or long cycles of repetitions.

To create coevalness, it is crucial to adopt the local conceptions that organize the past and relate them to "general social reality." This is not homochronism, but since it is grounded in local temporalities, it is not European-derived allochronism, either. It can often pose significant challenges, since such temporal representations do not form typical chronologies and adhere to mnemonic devices other than time lines. For instance, Basso's discussion of Apache morality tales in *Wisdom Sits in Places* (1996) weaves together multiple temporalities. Central to the book is the attachment of Apache stories about the past to places—"place-making is a way of constructing history itself" (1996, 6). The Apache stories of place names that Basso relates include multiple temporalities of stasis and change. For instance, the story of the place "Water Lies With Mud In An Open Container" (1996, 8) contains a narrative of change that discusses how a long time ago people traveled to the spot and settled there, but it also culminates in a statement of stasis: "Water and mud together, just as they were when our ancestors came here" (1996, 12). In addition to representing tales in which multiple Apache temporal frames interact, Basso offers a discussion of how the Apache organize their past with the major distinctions of "in the beginning," "long ago" (before the whitemen came), and "modern times" (since the whitemen came) (1996, 49–51). He describes how each temporal frame is linked to a different purpose with myths about the beginning of time serving "enlightenment and instruction"; historical tales of long ago used to criticize and edify behavior; and stories of modern times used to provide entertainment. The different historical modes convey a relationship of change and stasis that makes the past meaningful to the present, and Basso's temporal representations lead readers through his experience of discovering the Apache past in the present in his temporal frames.

There are also works that manage an interactive sense of coevalness by linking academic history and ethnographically revealed conceptions of the past. Mintz's *Worker in the Cane* (1974) does this. Each chapter title provides two temporal frames: a stage in Taso's life and a range of dates, for example, "Manhood: The Early Years (1927–39)" (1974, 99). Each chapter contains Taso's narratives followed by a brief section in which Mintz relates Taso's stories to dates and the political economy of Puerto Rico for the period. So, in the chapter just mentioned, Taso and his wife, Elisabeth, tell multiple stories with the most common temporal frames being "at that time" or "after that time" related to occasional dates. At the end of the chapter, Mintz provides chronology in which each significant event is assigned a date, for example, "Taso's blacklisting and his subsequent political disillusionment (1932–33)" (1974, 169). After presenting this chronology, Mintz offers general statements to provide additional historical context to Taso's life, such as: "By the 1930s most of the land on the south coast was in the hands of a few corporations, most of them North American. The political situation was very dark, and workers were forced to vote in line with their employers' interests to keep their homes and jobs" (1974, 169). Taso's life story and Mintz's history of Puerto Rico alternate in a way that enriches both narratives.

Richard Price's works on the Saramaka (1983, 1990) also represent local temporalities in contrapuntal relationship to conventional academic historical narratives. For instance, in *Alabi's World* (1990), he includes narratives of the same set of events from the perspectives of the Saramaka, the German Moravian missionaries, the Dutch colonial officials, and his own voice as an "ethnographic historian" (1990, xi–xvi). Each voice is identified with a different typeface in the book. In so doing, the implicit temporal principles that organize each narrative are juxtaposed, and Saramaka tales of the past become intelligible to readers familiar with European historical tropes because of the relationship of Saramaka temporal frames to familiar European frames. The clock-and-date consciousness of the colonial officials subtly contrasts with the date-and-time-of-day consciousness of the missionaries and the strong plots conveyed in Saramaka histories, and Price coordinates these different sources on the past. The dissonance between the European voices of the Moravian missionaries, the colonial officials, and the narratives of Saramaka is evident, and their lack of coevalness as a result of having different pasts is a source of ethnographic insight.

In these examples, the ethnography includes the writers' struggles to come to terms with the local temporality in relationship to temporalities familiar to readers of ethnographies, rather than obscuring local temporalities in the imposition of a European-derived chronology.

Thinking with Other Temporalities

I began with Derrida's favoring the naturalization of signified time in analyzing a story by Baudelaire. I end with an emphasis on how the choice of time is not just a matter of representation but is also a means of obtaining insight. Since temporalities shape research and analysis, adopting alternative temporalities tends to produce new and different questions. In contrast, naturalizing time tends to constrict knowledge and steers scholarship toward merely reproducing itself. A conscious and explicit representation of time can be deployed as a useful and creative tool for thought.

Colin Pittendrigh's work on mosquitos discussed in Chapter 2, which led to a fruitful career in chronobiology, is an example. Pittendrigh used the clock as a research tool to indicate when nets should be dropped on the mosquito-laden donkey, and to document the number and variety of mosquitos caught at each time. His research led to a conclusion that mosquito-biting behaviors were not tied to a particular clock time, but to humidity levels. Mosquito temporalities are structured by cycles of moisture—a humidity-driven temporality.

In another example, Michelle Bastian (2012) offers insights into climate change by postulating a "turtle clock." This temporality is constituted by the life span and changing populations of leatherback turtles. It defies the homochronic logic of universalizing time, and instead reveals the complex relationship of turtle temporalities with other features of the world, whether it be the ecological impact of shrimping techniques, or the slow response of humans to environmental changes. The temporality of turtles highlights that environmental policy is not a matter of annual budgets or five-year plans, but of living things in an unstable and unpredictable world.

Finally, my own work on roosters in medieval Europe revealed context-based temporalities (Birth 2011)—that roosters provide a good means of reckoning time during the predawn period, but not afterward. In effect, different temporalities can apply to different times of day rather than our current standard, which is a universalizing, homogenizing timescale that is equally imposable on anything.

In effect, the world is full of cycles, rhythms, sequences, and tempos that can be used to overcome the conceptual blinders that uniform timekeeping places on thought. The world is also full of cultural elaborations and interpretations of these cycles, rhythms, and sequences that can be used to challenge the European Enlightenment's grip on many approaches to human diversity.

REFERENCES

Abbott, Andrew. 2001. *Time Matters: On Theory and Method*. Chicago: University of Chicago Press.
Abram, Simone. 2014. The Time It Takes: Temporalities of Planning. *Journal of the Royal Anthropological Institute* 20(suppl.): 129–147.
Adam, Barbara. 1995. *Timewatch: The Social Analysis of Time*. Cambridge: Polity Press.
———. 1998. *Timescapes of Modernity: The Environment and Invisible Hazards*. London: Routledge.
———. 2010. History of the Future: Paradoxes and Challenges. *Rethinking History* 14(3): 361–378.
Addison, Joseph. 1744. *The Spectator* 6:328. London: Printed for J. and R. Tonson in the Strand. [Eighteenth-Century Collections Online. Gale. CUNY - Queens College. Accessed 29 April 2015].
Alkon, Paul. 1979. *Defoe and Fictional Time*. Athens: University of Georgia Press.
Allen, Thomas M. 2008. *A Republic in Time: Temporality and Social Imagination in Nineteenth-Century America*. Chapel Hill, NC: University of North Carolina Press.
Alleyne, Richard. 2011. Amazonian Tribe Has No Calendar and No Concept of Time. *Daily Telegraph*, May 20, 2011. http://www.telegraph.co.uk/news/science/science-news/8526287/Amazonian-tribe-has-no-calendar-and-no-concept-of-time.html
Anderson, Benedict. 2006. *Imagined Communities*. Rev. ed. London: Verso.
Appian. 1912. *Roman History*, vol. I. Ed. and trans. Brian McGing. Loeb Classical Library 2. Cambridge, MA: Harvard University Press.
Arias, E.F., G. Panfilo, and G. Petit. 2011. Timescales at the BIPM. *Metrologia* 48: S145–S153.

Aristotle. 1926. *Art of Rhetoric*. Trans. J.H. Freese. Loeb Classical Library 193. Cambridge, MA: Harvard University Press.

———. 1936. *Physics*. Ed. and trans. W.D. Ross. Oxford: Oxford University Press.

Asad, Talal. 2003. *Formations of the Secular: Christianity, Islam, Modernity*. Stanford: Stanford University Press.

Aschoff, Jürgen. 1960. Exogenous and Endogenous Components in Circadian Rhythms. In *Cold Spring Harbor Symposia on Quantitative Biology, Vol. 25: Biological Clocks*, 11–27. Cold Spring Harbor, NY: The Biological Laboratory.

Ashkanasy, Neal, Vipin Gupta, Melinda Mayfield, and Edwin Trevor-Roberts. 2004. Future Orientation. In *Culture, Leadership, and Organizations*, ed. Robert J. House, 282–342. Thousand Oaks: Sage.

Augustine. 1997. *The Confessions*. Trans. Maria Boulding. New York: Vintage.

Austen, Leo. 1939. The Seasonal Gardening Calendar of Kiriwina, Trobriand Islands. *Oceania* 9: 237–253.

Bachelard, Gaston. 2000. *The Dialectic of Duration*. Trans. Mary McAllester Jones. Manchester: Clinamen Press.

Bacon, Roger. 1859. *Opera quædamhactenus inedita. Vol. I. containing I.–Opus tertium. II.–Opus minus. III.–Compendium philosophiæ*. Ed. J.S. Brewer. London: Longman, Green, Longman, and Roberts.

Bakhtin, Mikhail. 1981. *The Dialogic Imagination*. Ed. and trans. Caryl Emerson and Michael Holquist. Austin: University of Texas Press.

Barsalou, L.W. 1988. The Content and Organization of Autobiographical Memories. In *Remembering Reconsidered*, eds. U. Neisser, and E. Winograd, 193–243. Cambridge: Cambridge University Press.

Bartky, Ian. 2007. *One Time Fits All: The Campaigns for Global Uniformity*. Stanford: Stanford University Press.

Bartlett, F. 1932. *Remembering*. Cambridge: Cambridge University Press.

Basso, Keith. 1996. *Wisdom Sits in Places*. Albuquerque: University of New Mexico Press.

Bastian, Adolf. 1860. *Der Mensch in der Geschichte zur Begründung einer Psychologischen Weltanschauung*, vol. 1. Leipzig: Otto Wigand.

Bastian, Michelle. 2012. Fatally Confused: Telling the Time in the Midst of Ecological Crises. *Journal of Environmental Philosophy* 9(1): 23–48.

Bear, Laura. 2014a. Doubt, Conflict, Mediation: The Anthropology of Modern Time. *Journal of the Royal Anthropological Institute* 20(suppl.): 3–30.

———. 2014b. Capital and Time: Uncertainty and Qualitative Measures of Inequality. *The British Journal of Sociology* 65(4): 639–649.

Becker, Gary S. 1965. A Theory of the Allocation of Time. *The Economic Journal* 75(299): 493–517.

Bede. 1844–1864. Liber de Vita et Miraculis Sancti Cuthberti, Lindisfarnensis Episcopi. In *Patrologia Latina*, vol. 94, ed. J.-P. Migne, columns 735–790. Turnhold: Typographi Brepols Editores Pontifici.

———. 1930. *Ecclesiastical History*. Trans. J.E. King. Cambridge, MA: Loeb Classical Library.
Behar, Ruth. 1986. *Santa Maria del Monte*. Princeton: Princeton University Press.
Benítez-Rojo, Antonio. 1996. *The Repeating Island*. Durham: Duke University Press.
Benjamin, Walter. 1968. Theses on the Philosophy of History. In *Illuminations*, ed. Hannah Arendt and trans. Harry Zohn, 253–264. New York: Schocken Books.
Berlin, Isaiah. 1976. *Vico and Herder*. New York: Viking.
Bernstein, Basil. 1971. *Class, Codes and Control*. London: Routledge and Kegan Paul.
Berntsen, Dorthe. 2001. Involuntary Memories of Emotional Events: Do Memories of Traumas and Extremely Happy Events Differ? *Applied Cognitive Psychology* 15(7): S135–S158.
Berthoud, Jacques. 1987. Mechanical Time in Eighteenth-Century English Literature. In *Science and Imagination in XVIIIth-Century British Culture*, ed. Sergio Rossi, 35–47. Milan: Edizioni Unicopli.
Bhabha, Homi. 1994. *The Location of Culture*. New York: Routledge.
Birth, Kevin K. 1994. Bakrnal: Coup, Carnival, and Calypso in Trinidad. *Ethnology* 33: 165–177.
———. 1999. *Any Time Is Trinidad Time*. Gainesville: University Press of Florida.
———. 2004. Finding Time: Studying the Concepts of Time Used in Daily Life. *Field Methods* 16(1): 70–84.
———. 2006. Past Times: Temporal Structure of History and Memory. *Ethos* 34: 192–210.
———. 2007. Time and the Biological Consequences of Globalization. *Current Anthropology* 48(2): 215–236.
———. 2008. *Bacchanalian Sentiments: Musical Experiences and Political Counterpoints in Trinidad*. Durham: Duke University Press.
———. 2011. The Regular Sound of the Cock: Context-Dependent Time Reckoning in the Middle Ages. *Kronoscope* 11(1–2): 125–144.
———. 2012. *Objects of Time: How Things Shape Temporality*. New York: Palgrave MacMillan.
———. 2013. Calendars: Representational Homogeneity and Heterogenous Time. *Time and Society* 22(2): 216–236.
———. 2014a. Non-Clocklike Features of Psychological Timing and Alternatives to the Clock Metaphor. *Timing and Time Perception* 2(3): 312–324.
———. 2014b. The Vindolanda Timepiece: Time and Calendar Reckoning in Roman Britain. *Oxford Journal of Archaeology* 33(4): 395–411.
Blackburn, Bonnie, and Leofranc Holford-Stevens. 1999. *The Oxford Companion to the Year*. Oxford: Oxford University Press.
Bloch, Ernst. 1986. *The Principle of Hope*. Trans. Neville Plaice, Stephen Plaice, and Paul Knight. Cambridge, MA: MIT Press.

———. 2000. *The Spirit of Utopia*. Stanford: Stanford University Press.
Bloch, Maurice. 1977. The Past and the Present in the Present. *Man* 12(2): 278–292.
———. 1979. Knowing the World or Hiding It. *Man* 14: 165–167.
———. 1998. *How We Think They Think*. Boulder, CO: Westview Press.
Borneman, John. 1998. *Subversions of International Order: Studies in the Political Anthropology of Culture*. Albany: State University of New York Press.
Borst, Arno. 1993. *The Ordering of Time*. Trans. Andrew Winnard. Chicago: University of Chicago Press.
Bourdieu, Pierre. 1963. The Attitude of the Algerian Peasant toward Time. In *Mediterranean Countrymen*, ed. Julian Pitt-Rivers, 55–72. Paris: Mouton.
———. 1979. *Algeria 1960*. Cambridge: Cambridge University Press.
Bourdillon, M.F.C. 1978. Knowing the World or Hiding It: A Response to Maurice Bloch. *Man* 13: 591–599.
Boyer, George, and Robert S. Smith. 2001. The Development of the Neoclassical Tradition in Labor Economics. *Industrial & Labor Relations Review* 54(2): 199–223.
Braudel, Fernand. 1980. *On History*. Trans. Sarah Matthews. Chicago: University of Chicago Press.
Brown, Gordon D.A., and Nick Chater. 2001. The Chronological Organization of Memory: Common Psychological Foundations for Remembering and Timing. In *Time and Memory*, eds. Christoph Hoerl, and Teresa McCormack, 77–110. Oxford: Oxford University Press.
Brown, Norman, Steven Shevell, and Lance J. Rips. 1986. Public Memories and Their Personal Context. In *Autobiographical Memory*, ed. David C. Rubin, 137–158. Cambridge: Cambridge University Press.
Bulhan, Hussein Abdilahi. 1985. *Frantz Fanon and the Psychology of Oppression*. New York: Plenum.
Bünning, Erwin. 1960. Opening Address: Biological Clocks. In *Cold Spring Harbor Symposia on Quantitative Biology, Vol. 25: Biological Clocks*, 1–10. Cold Spring Harbor, NY: The Biological Laboratory.
Cadbury, Edward. 1914. Some Principles of Industrial Organization: The Case for and against Scientific Management. *The Sociological Review* 7(2): 99–117.
Carr, D. 1986. *Time, Narrative and History*. Bloomington: Indiana University Press.
Castells, Manuel. 2000. *The Rise of the Network Society*. 2nd ed. Oxford: Blackwell.
Chakrabarty, Dipesh. 1992. Postcoloniality and the Artifice of History: Who Speaks for "Indian" Pasts? *Representations* 37: 1–26.
———. 1997. The Time of History and the Times of Gods. In *Politics of Culture in the Shadow of Capital*, eds. Lisa Lowe, and David Lloyd, 35–60. Durham: Duke University Press.
———. 2000. *Provincializing Europe*. Princeton: Princeton University Press.

Chance, John. 1996. Mesoamerica's Ethnographic Past. *Ethnohistory* 43: 379–403.
Cheney, Christopher Robert. 1970. *Handbook of Dates for Students of English History*. London: Offices of the Royal Historical Society.
Clark, Andy. 2008. *Supersizing the Mind: Embodiment, Action, and Cognitive Extension*. Oxford: Oxford University Press.
Clifford, James. 1983. On Ethnographic Authority. *Representations* 2: 118–146.
Clifford, James, and George Marcus, eds. 1986. *Writing Culture*. Berkeley: University of California Press.
CNRS. 1950. *Constantes Fondamentales de l'Astronomie*. Paris: Colloques Internationaux du Centre National de la Recherche Scientifique.
Cole, Jennifer. 2001. *Forget Colonialism*. Berkeley: University of California Press.
———. 2006. Malagasy and Western Conceptions of Memory: Implications for Postcolonial Politics and the Study of Memory. *Ethos* 34: 211–243.
Collins, John F. 2015. *Revolt of the Saints: Memory and Redemption in the Twilight of Brazilian Racial Democracy*. Durham, NC: Duke University Press.
Condon, Richard G. 1983. *Inuit Behavior and Seasonal Change in the Canadian Arctic*. Ann Arbor, MI: UMI Research Press.
Conway, M.A. 1992. A Structural Model of Autobiographical Memory. In *Theoretical Perspectives on Autobiographical Memory*, eds. M.A. Conway, D.C. Rubin, H. Spinnler, and W.A. Wagenaar, 167–194. Dordrecht: Kluwer Academic Press.
Corbin, Alain. 1998. *Village Bells: Sound and Meaning in the Nineteenth-Century French Countryside*. Trans. Martin Thom. New York: Columbia University Press.
Craig, Susan. 1985. Political Patronage and Community Resistance: Village Councils in Trinidad and Tobago. In *Rural Development in the Caribbean*, ed. P.I. Gomes, 173–193. Kingston, Jamaica: Heinemann.
Crapanzano, Vincent. 1986. Hermes' Dilemma: The Masking of Subversion in Ethnographic Description. In *Writing Culture*, eds. James Clifford, and George Marcus, 51–76. Berkeley: University of California Press.
———. 2003. Reflections on Hope as a Category of Social and Psychological Analysis. *Cultural Anthropology* 18(1): 3–32.
———. 2007. Co-Futures. *American Ethnologist* 34(3): 422–425.
Dallmann, Robert, Steven A. Brown, and Frédéric Gachon. 2014. Chronopharmacology: New Insights and Therapeutic Implications. *Annual Review of Pharmacology and Toxicology* 54: 339–361.
Davis, J. 1991. *Times and Identities*. Oxford: Clarendon Press.
Declerq, Georges. 2000. *Anno Domini: The Origins of the Christian Era*. Turnhout: Brepols.
Deleuze, Gilles. 1994. *Difference and Repetition*. Trans. Paul Patton. New York: Columbia University Press.

Deleuze, Gilles, and Felix Guattari. 1987. *A Thousand Plateaus: Capitalism and Schizophrenia*. Trans. Brian Massumi. Minneapolis: University of Minnesota Press.

Derrida, Jacques. 1992. *Given Time: I. Counterfeit Money*. Trans. Peggy Kamufs. Chicago: University of Chicago Press.

Dershowitz, Nachum, and Edward Reingold. 1997. *Calendrical Calculations*. Cambridge: Cambridge University Press.

Dionysius Exiguus. 1844–1864. Epistola Dionysii de Ratione Paschae. In *Patrologia Latina*, vol. 67, ed. J.-P. Migne, columns 514–520. Turnhold: Typographi Brepols Editores Pontifici.

Dohrn van-Rossum, Gerhard. 1996. *The History of the Hour: Clocks and Modern Temporal Orders*. Trans. Thomas Dunlap. Chicago: University of Chicago Press.

Edensor, Tim. 2006. Reconsidering National Temporalities: Institutional Times, Everyday Routines, Serial Spaces, and Synchronicities. *European Journal of Social Theory* 9(4): 525–545.

Einstein, Albert. 1992. Space-Time. In *The Treasury of the Encyclopedaeia Britannica*, ed. Clifton Fadiman, 371–383. New York: Viking.

Elias, Norbert. 1992. *Time: An Essay*. Oxford: Basil Blackwell.

Evans-Pritchard, E.E. 1939. Nuer Time-Reckoning. *Africa* 12(2): 189–216.

———. 1940. *The Nuer*. Oxford: Oxford University Press.

Fabian, Johannes. 1985. Culture, Time and the Object of Anthropology. *Berkshire Review* 20: 7–23.

———. 2002. *Time and the Other*. New York: Columbia University Press.

Fairfax, Thomas. 1776. *The Memoirs of General Fairfax: Wherein Is Contained an Account of All His Sieges and Battles in the North of England; Especially the Battles of Leeds, Wakefield, Manchester, &c*. Leeds: J. Bowling. [Eighteenth-Century Collections Online. Gale. CUNY - Queens College. Accessed 28 May 2015].

Fanon, Frantz. 1963. *The Wretched of the Earth*. Trans. Constance Farrington. New York: Grove.

———. 1967. *Black Skin White Masks*. Trans. Charles Lam Markmann. New York: Grove.

Feeney, Denis. 2007. *Caesar's Calendar: Ancient Time and the Beginnings of History*. Berkeley: University of California Press.

Ferry, Elizabeth. 2005. *Not Ours Alone: Patrimony, Value, and Collectivity in Contemporary Mexico*. New York: Columbia University Press.

———. 2006. Memory as Wealth, History as Commerce: Uses of Patrimony in a Central Mexican City. *Ethos* 34(2): 297–324.

Fielding, Henry. 1963. *Tom Jones*. New York: New American Library.

Fitzgerald, Robert. 1988. *British Labour Management & Industrial Welfare, 1846–1939*. London: Croom Helm.

Forte, Maximilian. 2005. *Ruins of Absence, Presence of Caribs: (Post)Colonial Representations of Aboriginality in Trinidad and Tobago.* Gainseville, FL: University Press of Florida.
Foucault, Michel. 1977. *Discipline and Punish: The Birth of the Prison.* Trans. Alan Sheridan. London: Allen Lane.
———. 1984. Nietzsche, Genealogy, History. In *The Foucault Reader*, ed. Paul Rabinow, 76–100. New York: Pantheon Books.
Frazer, James G. 1889. Questions on the Manners, Customs, Religion, Superstitions, &c. of Uncivilized or Semi-Civilized Peoples. *Journal of the Anthropological Institute of Great Britain and Ireland* 18: 431–440.
———. 1961. *Adonis Arris Osiris: Studies in the History of Oriental Religion.* New Hyde Park: University Books.
Freeman, Karen. 1998. Unhappy Marriage of Plants. *New York Times*, September 15, Science Watch, p. 4.
Freilich, Morris. 1960. *Cultural Diversity Among Trinidadian Peasants.* Ph.D. Dissertation, Columbia University.
Freire-Marreco, Barbara W., and John Linton Myres. 1912. *Notes and Queries On Anthropology.* 4th ed. London: The Royal Anthropological Institute.
Friedman, Thomas. 2005. *The World Is Flat: A Brief History of the Twenty-First Century.* New York: Farrar, Straus and Giroux.
Galison, Peter. 1987. *How Experiments End.* Chicago: University of Chicago Press.
———. 2003. *Einstein's Clocks and Poincare's Maps: Empires of Time.* New York: W. W. Norton.
Geertz, Clifford. 1973. *The Interpretation of Cultures.* New York: Basic Books.
Gell, Alfred. 1992. *The Anthropology of Time: Cultural Constructions of Temporal Maps and Images.* Oxford: Berg.
Glennie, Paul, and Nigel Thrift. 2009. *Shaping the Day: A History of Timekeeping in England and Wales.* Oxford: Oxford University Press.
Glissant, Édouard. 1989. *Caribbean Discourse.* Trans. J. Michael Dash. Charlottesville: University of Virginia Press.
Gompf, Heinrich S., Patrick M. Fuller, Samer Hattar, Clifford B. Saper, and Jun Lu. 2015. Impaired Circadian Photosensitivity in Mice Lacking Glutamate Transmission from Retinal Melanopsin Cells. *Journal of Biological Rhythms* 30(1): 35–41.
Goodman, Jane. 2003. The Proverbial Bourdieu: Habitus and the Politics of Representation in the Ethnography of Kabylia. *American Anthropologist* 105: 782–793.
Gore, Al. 2006. *An Inconvenient Truth: The Planetary Emergency of Global Warming and What We Can Do About It.* New York: Rodale.
Greenhouse, Carol J. 1996. *A Moment's Notice: Time Politics across Culture.* Ithaca, NY: Cornell University Press.
Gurevich, A.J. 1976. Time as a Problem of Cultural Hsitory. In *Cultures and Time*, ed. Louis Gardet, 229–245. Paris: The UNESCO Press.

Guyer, Jane. 2007. Prophecy and the Near Future: Thoughts on Macroeconomic, Evangelical, and Punctuated Time. *American Ethnologist* 34(3): 409–421.
Habermas, Jürgen. 1971. *Toward a Rational Society: Student Protest, Science and Politics*. Trans. Jeremy J. Shapiro. Boston: Beacon.
Halbwachs, Maurice. 1992. *On Collective Memory*. Trans. Lewis Coser. Chicago: University of Chicago Press.
Hallowell, A. Irving. 1955. Temporal Orientation in Western Civilization and in a Preliterate Society. In *Culture and Experience*, 216–235. Philadelphia: University of Pennsylvania Press.
Hamilton, Robert. 1787. *The Duties of a Regimental Surgeon Considered*, vol. 2. London: J. Johnson. [Eighteenth-Century Collections Online. Gale. CUNY - Queens College. Accessed 1 June 2015].
Harris, Wilson. 1999. *Selected Essays of Wilson Harris*. Ed. Andrew Bundy. New York: Routledge.
Harvey, David. 1989. *The Condition of Postmodernity: An Enquiry into the Origins of Culture Change*. Oxford: Blackwell.
———. 1990. Between Space and Time: Reflections on the Geographical Imagination. *Annals of the Association of American Geographers* 80: 418–434.
———. 1993. From Space to Place and Back Again: Reflections on the Condition of Postmodernity. In *Mapping the Futures: Local Cultures, Global Change*, eds. Jon Bird, Barry Curtis, Tim Putnam, George Robertson, and Lisa Tickner, 3–29. London: Routledge.
Hassan, Robert. 2003. *The Chronoscopic Society: Globalization, Time and Knowledge in the Network Society*. New York: Peter Lang.
Hegel, G.W.F. 1977. *Phenomenology of Spirit*. Trans. A.V. Miller. Oxford: Clarendon.
Heidegger, Martin. 1962. *Being and Time*. Trans. John Macquarrie and Edward Robinson. New York: Harper and Row.
———. 1989. *The Concept of Time*. Trans. William McNeill. Oxford: Blackwell.
———. 1992. *History of the Concept of Time*. Trans. Theodore Kisiel. Bloomington: Indiana University Press.
Henzi, S.P., R.W. Byrne, and A. Whiten. 1992. Patterns of Movement by Baboons in the Drakensberg Mountains: Primary Responses to the Environment. *International Journal of Primatology* 13(6): 601–629.
Hesiod. 1988. *Works and Days*. Trans. M.L. West. Oxford: Oxford University Press.
Hintzen, Percy C. 1989. *The Costs of Regime Survival: Racial Mobilization, Elite Domination, and Control of the State in Guyana and Trinidad*. Cambridge: Cambridge University Press.
Hobsbawm, Eric. 1997. *On History*. New York: The New Press.
Hobsbawm, Eric, and Terence Ranger, eds. 1983. *The Invention of Tradition*. Cambridge: Cambridge University Press.
Hodder, Ian. 2003. Archaeological Reflexivity and the "Local" Voice. *Anthropological Quarterly* 76: 55–69.

Hodges, Matt. 2014. Immanent Anthropology: A Comparative Study of 'Process' in Contemporary France. *Journal of the Royal Anthropological Institute* 20(suppl.): 33–51.
Hoskins, J. 1997. *The Play of Time*. Berkeley: University of California Press.
Howe, Leopold E.A. 1981. The Social Determination of Knowledge: Maurice Bloch and Balinese Time. *Man* (N.S.) 16: 220–234.
Hubert, Henri. 1999. *Essay on Time: A Brief Study of the Representation of Time in Religion and Magic*. Trans. Robert Parkin and Jacqueline Redding. Oxford: Durkheim Press.
Hume, Lynne. 2004. Accessing the Eternal: Dreaming 'The Dreaming' and Ceremonial Performance. *Zygon* 39: 237–258.
Hunnicutt, Benjamin K. 1996. *Kellogg's Six Hour Day*. Philadelphia: Temple University Press.
Hutton, Ronald. 1996. *The Stations of the Sun: A History of the Ritual Year in Britain*. Oxford: Oxford University Press.
Huygens, Christian. 1673. *Horologium Oscillatorium*. Paris: F. Muguet.
Isidore of Seville. 2005. *Etymologies*. Trans. Priscilla Throop. Charlotte, VT: MedievalMS.
James, C.L.R. 1980. *Notes on Dialectics: Hegel, Marx, Lenin*. Westport, CT: Lawrence Hill.
——— 1984. *Party Politics in the West Indies*. Port of Spain: Inprint.
James, Jason. 2006. Undoing Trauma: Reconstructing the Church of Our Lady in Dresden. *Ethos* 34(2): 244–272.
———. 2012. *Preservation and National Belonging in Eastern Germany: Heritage Fetishism and Redeeming Germanness*. New York: Palgrave Macmillan.
James, William. 1996. *The Pluralistic Universe*. Lincoln: University of Nebraska Press.
Jameson, Fredric. 1991. *Postmodernism or, the Cultural Logic of Late Capitalism*. Durham: Duke University Press.
Jarjisian, Stephen G., Matthew P. Butler, Matthew J. Paul, Ned J. Place, Brian J. Prendergast, Lance J. Kriegsfeld, and Irving Zucker. 2015. Dorsomedial Hypothalamic Lesions Counteract Decreases in Locomotor Activity in Male Syrian Hamsters Transferred from Long to Short Day Lengths. *Journal of Biological Rhythms* 15(30): 42–52.
Jemmat, Catherine. 1765. *The Memoirs of Mrs. Catherine Jemmat, Daughter of the Late Admiral Yeo, of Plymouth, Written by Herself*. London: Charing-Cross. [Eighteenth-Century Collections Online. Gale. CUNY - Queens College. Accessed 1 June 2015].
Johnson, Maynard S. 1939. Effect of Continuous Light on Periodic Spontaneous Activity of White-Footed Mice (*Peromyscus*). *Journal of Experimental Zoology* 82(2): 314–328.
Johnson, Samuel. 1779. *The Works of the English Poets. With Prefaces, Biographical and Critical*, vol. 32. London: H. Hughs. [Eighteenth-Century Collections Online. Gale. CUNY - Queens College. Accessed 1 June 2015].

Kanigel, Robert. 1997. *One Best Way: Frederic Winslow Taylor and the Enigma of Efficiency*. New York: Viking.
Kant, Immanuel. 2003. *Critique of Pure Reason*. Trans. Norman Kemp Smith. New York: Palgrave MacMillan.
Kelly, John D. 1998. Time and the Global: Against the Homogeneous, Empty Communities in Contemporary Social Theory. *Development and Change* 29: 839–871.
Kermode, Frank. 2000. *The Sense of an Ending*. New York: Oxford University Press.
Kinneavy, James L. 2002. *Kairos* in Classical and Modern Rhetorical Theory. In *Rhetoric and Kairos: Essays in History Theory, and Praxis*, eds. Phillip Sipiora, and James S. Baumlin, 58–76. Albany: State University of New York Press.
Kinneavy, James L., and Catherine R. Eskin. 1994. *Kairos* in Aristotle's *Rhetoric*. *Written Communication* 11: 131–142.
Koepping, Klaus-Peter. 1983. *Adolf Bastian and the Psychic Unity of Mankind: The Foundations of Anthropology in Nineteenth Century Germany*. St. Lucia: University of Queensland Press.
Koselleck, Reinhart. 1985. *Futures Past*. Cambridge, MA: MIT Press.
Kuhn, Thomas S. 1970. *The Structure of Scientific Revolutions*. Chicago: University of Chicago Press.
Kurbat, M.A., S.K. Shevell, and L.J. Rips. 1998. A Year's Memories: The Calendar Effect in Autobiographical Recall. *Memory and Cognition* 26: 532–552.
Lakoff, Andrew. 2008. The Generic Biothreat, or, How We Became Unprepared. *Cultural Anthropology* 23(3): 399–428.
Landes, D. 1983. *Revolution in Time*. Cambridge, MA: Belknap Press.
Latour, Bruno. 2004. *The Politics of Nature: How to Bring the Sciences into Democracy*. Cambridge, MA: Harvard University Press.
Latour, Bruno, and Steven Woolgar. 1986. *Laboratory Life: The Construction of Scientific Facts*. Princeton: Princeton University Press.
Lazar, Sian. 2014. Historical Narrative, Mundane Political Time, and Revolutionary Moments: Coexisting Temporalities in the Lived Experience of Social Movements. *Journal of the Royal Anthropological Institute* 20(suppl.): 91–108.
Leach, Edmund. 1950. Primitive Calendars. *Oceania* 20: 245–262.
Lederman, Rena. 1990. Pretexts for Ethnography: On Reading Fieldnotes. In *Fieldnotes: The Makings of Anthropology*, ed. Roger Sanjek, 71–91. Ithaca: Cornell University Press.
Leliavski, Alexei, Rebecca Dumbell, Volker Ott, and Henrik Oster. 2015. Adrenal Clocks and the Role of Adrenal Hormones in the Regulation of Circadian Physiology. *Journal of Biological Rhythms* 30(1): 20–34.
Levine, Judah. 2001. GPS and the Legal Traceability of Time. *GPS World* 12(1): 52–57.
Levine, Robert. 1997. *A Geography of Time: The Temporal Misadventures of a Social Psychologist*. New York: Basic Books.

Levitas, Ruth. 2013. *Utopia as Method: The Imaginary Reconstitution of Society.* New York: Palgrave Macmillan.
Lewis, H. Gregg. 1957. Hours of Work and Hours of Leisure. In *Proceedings of the Industrial Relations Research Association*, 196–207. Princeton University.
Lewis, W. Arthur. 1954. Economic Development with Unlimited Supplies of Labour. *Manchester School* 22(2): 139–191.
———. 1955. *The Theory of Economic Growth.* Homewood, IL: Richard D. Irwin.
Limbert, Mandana. 2010. *In the Time of Oil.* Stanford: Stanford University Press.
Livingstone, Frank B. 1958. Anthropological Implications of Sickle Cell Gene Distribution in West Africa. *American Anthropologist* 60(3): 533–562.
Luhmann, Niklas. 1998. *Observations on Modernity.* Palo Alto, CA: Stanford University Press.
Lyotard, Jean-François. 1984. *The Postmodern Condition.* Minneapolis: University of Minnesota Press.
Macey, Samuel L. 1980. *Clocks and the Cosmos: Time in Western Life and Thought.* Hamden, CT: Archon.
Malinowski, B. 1927. Lunar and Seasonal Calendar in the Trobriands. *The Journal of the Royal Anthropological Institute of Great Britain and Ireland* 57: 203–215.
———. 1961. *Argonauts of the Western Pacific.* New York: Dutton.
———. 1989. *A Diary in the Strict Sense of the Term.* Stanford: Stanford University Press.
Manoogian, Emily N.C., Tanya L. Leise, and Eric L. Bittman. 2015. Phase Resetting in Duper Hamsters: Specificity to Photic Zeitgebers and Circadian Phase. *Journal of Biological Rhythms* 30(2): 129–143.
Marcus, George, and Dick Cushman. 1982. Ethnographies as Texts. *Annual Review of Anthropology* 11: 25–69.
Marcuse, Herbert. 1968. *Negations: Essays in Critical Theory.* Trans. Jeremy J. Shapiro. Boston: Beacon.
Marramao, Giacomo. 2007. *Kairós: Towards and Ontology of "Due Time."* Aurora, CO: The Davies Group.
Marx, Karl. 1970. *A Contribution to the Critique of Political Economy.* Trans. S.W. Ryazanskaya. Moscow: Progress Publishers.
———. 1977. *Capital*, vol. 1. Trans. Ben Fowkes. New York: Vintage.
———. 1978. *Capital*, vol. 2. Trans. David Ferbach. London: Penguin.
McIntosh, Ian. 2000. *Aboriginal Reconciliation and the Dreaming.* Boston: Allyn and Bacon.
McLuhan, Marshall. 1994. *Understanding Media: The Extensions of Man.* Cambridge, MA: MIT Press.
Mead, George Herbert. 1932. *Philosophy of the Present.* Chicago: Open Court.
Menaker, Michael, Zachary C. Murphy, and Michael T. Sellix. 2013. Central Control of Peripheral Oscillators. *Current Opinion in Neurobiology* 23: 741–746.

Milham, Willis I. 1947. *Time and Timekeepers*. New York: Macmillan.
Miller, Daniel. 1994. *Modernity: An Ethnographic Approach: Dualism and Mass Consumption in Trinidad*. Oxford: Berg.
Mills, David L. 2011. *Computer Network Time Synchronization: The Network Time Protocol on Earth and in Space*. Boca Raton: CRC Press.
Mintz, Sidney. 1974. *Worker in the Cane*. New York: Norton.
———. 1985. *Sweetness and Power: The Place of Sugar in Modern History*. New York: Penguin.
———. 1993. Enduring Substances, Trying Theories: The Caribbean Region as Oikoumenê. *Journal of the Royal Anthropological Institute* 2: 289–311.
Miyazaki, Hirokazu. 2003. The Temporalities of the Market. *American Anthropologist* 105(2): 255–265.
———. 2004. *The Method of Hope: Anthropology, Philosophy, and Fijian Knowledge*. Palo Alto: Stanford University Press.
———. 2006. Economy of Dreams: Hope in Global Capitalism and Its Critiques. *Cultural Anthropology* 21(2): 147–172.
Monthly Magazine and British Register. 1798. Letter to the Editor. 6(35): 97–98.
Mulla, Sameena. 2014. *The Violence of Car: Rape Victims, Forensic Nurses, and Sexual Assault Intervention*. New York: New York University Press.
Mumford, Lewis. 1963. *Technics and Civilization*. New York: Harcourt, Brace and World.
Munn, Nancy. 1986. *The Fame of Gawa*. Durham, NC: Duke University Press.
———. 1992. The Cultural Anthropology of Time: A Critical Essay. *Annual Reviews of Anthropology* 21: 93–123.
Murakami, Kyoko, and David Middleton. 2006. Grave Matters: Emergent Networks and Summation in Remembering and Reconciliation. *Ethos* 34(2): 273–296.
Nelson, R.A., D.D. McCarthy, S. Malys, J. Jevine, B. Guinot, H.F. Fliegel, R.L. Beard, and T.R. Bartholmew. 2001. The Leap Second: Its History and Possible Future. *Metrologia* 38: 509–529.
Newton, Isaac. 1714. *Philosophiae Naturalis Principia Mathematica*. Amsterdam: Sumptibus Societatis. [Accessed from Bayerische Staatsbibliotech].
———. 1934. *Sir Isaac Newton's Mathematical Principles of Natural Philosophy and His System of the World*. Berkeley: University of California Press.
Nielsen, Morten. 2014. A Wedge of Time: Futures in the Present and Presents without Futures in Maputo, Mozambique. *Journal of the Royal Anthropological Institute* 20(suppl.): 166–182.
Nilsson, Martin Persson. 1920. *Primitive Time-Reckoning: A Study in the Origins and First Development of the Art of Counting Time Among the Primitive and Early Culture Peoples*. Lund: C.W.K. Gleerup.
Northcott, C.H. 1956. *Personnel Management: Principles and Practice*. New York: Philosophical Library.

Nowotny, Helga. 1994. *Time: The Modern and Postmodern Experience*. Trans. Neville Plaice. Cambridge: Polity Press.
Nugent, David. 2012. Democracy, Temporalities of Capitalism, and Dilemmas of Inclusion in Occupy Movements. *American Ethnologist* 39(2): 280–283.
Obeyesekere, Gananath. 1992. *The Apotheosis of Captain Cook*. Princeton: Princeton University Press.
Opitz, Sven, and Ute Tellmann. 2015. Future Emergencies: Temporal Politics in Law and Economy. *Theory, Culture and Society* 32(2): 107–129.
Paneth, N. 1993. Neurobehavioral Effects of Power-Frequency Electromagnetic Fields. *Environmental Health Perspectives* 101(suppl. 4): 101–106.
Peel, J.D.Y. 1984. Making History: The Past in the Ijesha Present. *Man* (N.S.) 19: 111–132.
Piketty, Thomas. 2014. *Capital in the Twenty-First Century*. Trans. Arthur Goldhammer. Cambridge, MA: The Belknap Press.
Pittendrigh, C.S. 1950. The Ecoclimatic Divergence of *Anopheles bellator* and *A. homunculus*. *Evolution* 4(1): 43–63.
———. 1954. On Temperature Independence in the Clock-system Controlling Emergence Time in *Drosophilia*. *Proceedings of the National Academy of Sciences* 40: 1018–1029.
———. 1960. Circadian Rhythms and the Circadian Organization of Living Systems. In *Cold Spring Harbor Symposia on Quantitative Biology, Vol. 25: Biological Clocks*, 159–184. Cold Spring Harbor, NY: The Biological Laboratory.
———. 1993. Temporal Organization: Reflections of a Darwinian Clock-Watcher. *Annual Review of Physiology* 55: 17–54.
Plato. 1914. *Euthyphro. Apology. Crito. Phaedo. Phaedrus*. Trans. Harold North Fowler. Loeb Classical Library 36. Cambridge, MA: Harvard University Press.
Pliny the Elder. 1938. *Natural History*, vol. I: Books 1–2. Trans. H. Rackham. Loeb Classical Library 330. Cambridge, MA: Harvard University Press.
Poe, Edgar Allan. 1975. *Complete Tales and Poems*. New York: Vintage.
Pope, Alexander. 1970. *An Essay on Criticism*. Menston, UK: The Scolar Press.
Postill, John. 2002. Clock and Calendar Time: A Missing Anthropological Problem. *Time and Society* 11(2/3): 251–270.
Postone, Moishe. 2003. *Time, Labor, and Social Domination: A Reinterpretation of Marx's Critical Theory*. Cambridge: Cambridge University Press.
Powdermaker, Hortense. 1966. *Stranger and Friend: The Way of an Anthropologist*. New York: W. W. Norton.
Price, Richard. 1983. *First-Time*. Baltimore: Johns Hopkins University Press.
———. 1990. *Alabi's World*. Baltimore: Johns Hopkins University Press.
———. 1998. *The Convict and the Colonel*. Boston: Beacon.
Rabinow, Paul. 1988. Beyond Ethnography: Anthropology as Nominalism. *Cultural Anthropology* 3: 355–364.

Rappaport, Joanne. 1998. *The Politics of Memory.* Cambridge: Cambridge University Press.
Rhys, John. 1892. *Lectures on the Origin and Growth of Religion as Illustrated by Celtic Heathendom.* London: Williams and Norgate.
Rice, Prudence. 2004. *Maya Political Science.* Austin: University of Texas Press.
Ricoeur, Paul. 1984. *Time and Narrative*, vol. 1. Trans. Kathleen McLaughlin and David Pellauer. Chicago: University of Chicago Press.
———. 1985. *Time and Narrative*, vol. 2. Trans. Kathleen McLaughlin and David Pellauer. Chicago: University of Chicago Press.
———. 1988. *Time and Narrative*, vol. 3. Trans. Kathleen Blamey and David Pellauer. Chicago: University of Chicago Press.
Ringel, Felix. 2014. Post-Industrial Times and the Unexpected: Endurance and Sustainability in Germany's Fastest-Shrinking City. *Journal of the Royal Anthropological Institute* 20(suppl.): 52–70.
Rivers, W.H.R. 1910. The Genealogical Method of Anthropological Inquiry. *The Sociological Review* 3(1): 1–12.
——— 1911. The Ethnological Analysis of Culture. *Science* 34(874): 385–397.
Robbins, Joel. 2007. Continuity Thinking and the Problem of Christian Culture: Belief, Time and the Anthropology of Christianity. *Current Anthropology* 48(1): 5–38.
Robinson, John A. 1986. Temporal Reference Systems and Autobiographical Memory. In *Autobiographical Memory*, ed. David C. Rubin, 159–188. Cambridge: Cambridge University Press.
Robinson, John A.T. 1968. *In the End God.* New York: Harper and Row.
Rodney, Walter. 1981. *A History of the Guyanese Working People, 1881–1905.* Baltimore: Johns Hopkins University Press.
Roediger, David R., and Philip S. Foner. 1989. *Our Own Time: A History of American Labor and the Working Day.* London: Verso.
Ryan, Selwyn. 1972. *Race and Nationalism in Trinidad and Tobago.* Toronto: University of Toronto Press.
Sahlins, Marshall. 1985. *Islands of History.* Chicago: University of Chicago Press.
Sanjek, Roger, ed. 1990. *Fieldnotes: The Makings of Anthropology.* Ithaca: Cornell University Press.
———, ed. 1991. The Ethnographic Present. *Man* (N.S.) 26: 609–628.
Schutz, Alfred. 1967. *The Phenomenology of the Social World.* Trans. George Walsh and Frederick Lehnert. Evanston: Northwestern University Press.
Scott, David. 2004. *Conscripts of Modernity: The Tragedy of Colonial Enlightenment.* Durham: Duke University Press.
Seaman, Rob. 2014. The Meaning of a Day. In *Requirements for UTC and Civil Timekeeping on Earth*, eds. John H. Seago, Robert L. Seaman, P. Kenneth Seidelmann, and Steven L. Allen. *American Astronautical Society Science and Technology Series* 115: 171–186.

Sellix, Michael T. 2015. Circadian Clock Function in the Mammalian Ovary. *Journal of Biological Rhythms* 30(1): 7–19.
Shaw, Matthew. 2011. *Time and the French Revolution: The Republican Calendar, 1789–Year XIV.* Royal Historical Society Studies in History, New Series. Woodbridge, Suffolk: Boydell Press.
Sherman, Stuart. 1996. *Telling Time: Clocks, Diaries, and English Diurnal Form, 1660–1785.* Chicago: University of Chicago Press.
Shum, Michael. 1998. The Role of Temporal Landmarks in Autobiographical Memory Processes. *Psychological Bulletin* 124(3): 423–442.
Singer, S. Fred, and Dennis T. Avery. 2008. *Unstoppable Global Warming: Every 1,500 Years.* Lanham, MD: Rowman and Littlefield.
Sinha, Chris. 2014a. Is Space-Time Metaphorical Mapping Universal? In *Multilingual Cognition and Language Use: Processing and Typological Perspectives*, eds. Luna Filipović, and Martin Pütz, 183–202. Amsterdam: John Benjamins.
———. 2014b. Living in the Model: The Cognitive Ecology of Time—A Comparative Study. *Model-Based Reasoning in Science and Technology. Studies in Applied Philosophy, Epistemology and Rational Ethics* 8: 55–73.
Sinha, Chris, Vera da Silva Sinha, Jörg Zinken, and Wany Sampaio. 2011. When Time Is Not Space: The Social and Linguistic Construction of Time Intervals and Temporal Event Relations in an Amazonian Culture. *Language and Cognition* 3(1): 137–169.
Smith, Adam. 1994. *The Wealth of Nations.* New York: Modern Library.
Smith, Mark M. 1997. *Mastered by the Clock.* Chapel Hill: University of North Carolina Press.
Stanner, W.E.H. 1979. The Dreaming. In *White Man Got No Dreaming*, 23–40. Canberra: Australian National University Press.
Stern, Pamela. 2003. Upside-Down and Backwards: Time Discipline in a Canadian Inuit Town. *Anthropologica* 45: 147–161.
Stern, Sacha. 2012. *Calendars in Antiquity: Empires, States, and Societies.* Oxford: Oxford University Press.
Stewart, Charles. 2012. *Dreaming and Historical Consciousness in Island Greece.* Cambridge: Harvard University Press.
Stiegler, Bernard. 1998. *Technics and Time: The Fault of Epimetheus.* Trans. Richard Beardsworth and George Collins. Stanford: Stanford University Press.
Taft, Robert. 1993. *The Liturgy of the Hours in East and West.* 2nd ed. Collegeville, MN: The Liturgical Press.
Taussig, Michael. 1980. *The Devil and Commodity Fetishism in South America.* Chapel Hill: University of North Carolina Press.
Taylor, Charles. 2007. *A Secular Age.* Cambridge, MA: Belknap.
Taylor, Frederick Winslow. 1911. *The Principles of Scientific Management.* New York: Harper and Brothers.

Thompson, E.P. 1967. Time, Work-discipline, and Industrial Capitalism. *Past and Present* 38: 56–97.
Tignor, Robert. 2004. Unlimited Supplies of Labor. *The Manchester School* 72(6): 691–711.
Tillich, Paul. 1948. *The Protestant Era*. Trans. James Luther Adams. Chicago: University of Chicago Press.
Trouillot, Michel-Rolph. 1995. *Silencing the Past*. Boston: Beacon.
Traweek, Sharon. 1988. *Beamtimes and Lifetimes: The World of High Energy Physicists*. Cambridge: Harvard University Press.
Turton, David, and Clive Ruggles. 1978. Agreeing to Disagree: The Measurement of Duration in a Southwestern Ethiopian Community. *Current Anthropology* 19(3): 585–600.
U.S. Department of Defense. 2014. *Climate Change Adaptation Roadmap*. Alexandria, VA: Office of the Deputy Under Secretary of Defense for Installations and Environment (Science & Technology Directorate).
Valéry, Paul. 1973. *Cahiers*, vol. 1. Ed. Judith Robinson. Paris: Gallimard.
Vegetius. 1967. *Epitoma Rei Militaris*. Ed. Carolus Lang. Stuttgart: B. G. Teubner.
Vélez-Ibáñez, Carlos. 1996. *Border Visions: Mexican Cultures of the Southwest United States*. Tucson: University of Arizona Press.
Verne, Jules. 1995. *Around the World in Eighty Days*. Trans. William Butcher. Oxford: Oxford University Press.
Vico, Giambattista. 1984. *The New Science*. Trans. Thomas Goddard Bergin and Max Harold Fisch. Ithaca, NY: Cornell University Press.
Virilio, Paul. 2010. *The Futurism of the Instant: Stop-Eject*. Trans. Julie Rose. Cambridge: Polity Press.
Wackermann, Jiří. 2008. Measure of Time: A Meeting Point of Psaychophysics and Fundamental Physics. *Mind and Matter* 6(1): 9–50.
———. 2011. On Clocks, Models and Metaphors: Understanding the Klepsydra Model. In *Multidisciplinary Aspects of Time Perception*, eds. Argiro Vatakis, Anna Esposito, Maria Giagkou, Fred Cummins, and Georgios Papadelis, 246–257. Berlin: Springer-Verlag.
Wackermann, Jiří, and Werner Ehm. 2006. The Dual Klepsydra Model of Internal Time Representation and Time Reproduction. *Journal of Theoretical Biology* 239: 482–493.
Warman, G.R., H.M. Tripp, V.L. Warman, and J. Arendt. 2003. Circadian Neuroendocrine Physiology and Electromagnetic Field Studies: Precautions and Complexities. *Radiation Protection Dosimetry* 106(4): 369–373.
West-Pavlov, Russell. 2013. *Temporalities*. London: Routledge.
Wever, Rütger A. 1979. *The Circadian System of Man: Results of Experiments Under Temporal Isolation*. New York: Springer-Verlag.

White, Geoffrey. 1999. Emotional Remembering: The Pragmatics of National Memory. *Ethos* 27(4): 505–529.
White, Hayden. 1975. *Metahistory: The Historical Imagination in Nineteenth-Century Europe*. Baltimore: Johns Hopkins University Press.
Wickman, Matthew. 2007. *The Ruins of Experience: Scotland's "Romantic" Highlands and the Birth of the Modern Witness*. Philadelphia: University of Pennsylvania Press.
Wilcox, Donald J. 1987. *The Measure of Times Past: Pre-Newtonian Chronologies and the Rhetoric of Relative Time*. Chicago: University of Chicago Press.
Wynter, Sylvia. 1984. The Ceremony Must Be Found: After Humanism. *Boundary 2* 12(3): 19–70.
Zerubavel, Eviatar. 1977. The French Republican Calendar: A Case Study in the Sociology of Time. *American Sociological Review* 42(6): 868–877.
———. 1979. *Patterns of Hospital Life*. Chicago: University of Chicago Press.
———. 2003. *Time Maps: Collective Memory and the Social Shape of the Past*. Chicago: University of Chicago Press.

Index

A

Abbott, Andrew, 2
Abram, Simone, 102
absolute time, 15, 34. *See also* Newton, Isaac
accuracy, 7, 8, 10, 11, 13, 15, 22, 23, 29, 77, 80, 83, 91, 118
Adam, Barbara, 1, 34, 44, 93
Addison, Joseph, 12
age, 64, 76–7, 84–8. *See also* life stage
agriculture, 7, 65, 67, 83, 104, 106, 126
Aidan, Saint, 25
algorithms, 6, 26, 29, 30, 34, 65, 119
Alkon, Paul, 72–3
Allen, Thomas, 28
Alleyne, Richard, 20
allochronism, 28, 121–2, 124–5, 126, 131, 135, 138
Amazonia, 20–1
Amondawa, 20–1
Anderson, Benedict, 15, 25, 68–9, 71, 72, 100–1, 110, 113
anthropology, 17–20, 22, 28, 94, 117–18, 123–5, 128, 131–40
  assumptions of, 18–20, 22, 136
  biases in, 20, 28, 117–18, 133–4

"Any time is Trinidad time", 7, 60, 67, 74, 77, 119–20, 121
Appian, 51
arbitrariness of clock/calendar time, 9, 14, 69, 115
arbitrariness of signs, 8, 69
Arias, Felicitas, 6
Aristotle, 49, 136
arithmetic, 26
artificiality, x, 8, 14, 23–4, 118
Asad, Talal, 15, 30, 71
Aschoff, Jürgen, 38
Ashkanasy, Neal, 129
astronomy, 5, 8, 13–14, 20, 22, 26, 118
atomic time. *See* clock—atomic
attempted coup d'état of 1990, 99–101, 109–13
Augustine of Hippo, Saint, 79, 117, 135–6
Austen, Leo, 19–20
average, 5–6, 33, 48–9, 58–61
average labor time, 48, 50–3, 56, 58–63, 66–8

B

baboons, 33, 48. See also primatology
Babylonia, xi
Bachelard, Gaston, 95–6
Bacon, Roger, 118
Bakhtin, Mikhail, 71, 73, 124, 132
Barsalou, L. W., 73
Bartky, Ian, 116
Bartlett, F., 86
Basso, Keith, 138
Bastian, Adolf, x
Bastian, Michelle, 140
Baudelaire, Charles, 4–12, 14–16, 48, 73, 140
BC/AD, 29
BCE/CE, 29
Bear, Laura, 22, 68
Becker, Gary, 53, 67
Bede, 25, 135
beginnings, 72, 134
Behar, Ruth, 125
bells, 9–10, 14
Bénitez-Rojo, Antonio, x, 53
Benjamin, Walter, 15, 69, 71
Berlin, Isaiah, 24
Bernstein, Basil, 122
Berntsen, Dorthe, 73
Berthoud, Jacques, 30
Bhabha, Homi, 27, 69, 89–90, 101
"biological clock", 36–8, 42–4
biology, xi–xii, 34, 37, 41–2, 44–5. See also chronobiology
BIPM. See International Bureau of Weights and Measures
birth of Jesus, 29, 135
Blackburn, Bonnie, 29
Bloch, Ernst, ix, 94
Bloch, Maurice, 21–2, 89, 134
Borneman, John, 125
Borst, Arno, 138
Bourdeiu, Pierre, 129
Bourdillon, M. F. C., 21
Boyer, George, 53
Braudel, Fernand, 72
Brown, Gordon, 73
Brown, Norman, 73
Bulhan, Hussein Abdilahi, 89
bull grass, 36, 56–7
Bünning, Erwin, 36–7

C

Cadbury, Edward, 48
Cadbury company, 48
calendar time, x–xi, 3, 21, 24, 27–8, 34, 48, 69, 115
calendars, 1, 9, 17–22, 26–30, 73–4, 99, 118–19, 138
 Celtic, 22
 Chinese, 26–7
 Gregorian, xi, 3, 18–20, 22–3, 26–30
 Hindu, 23, 26–7
 Jewish, 20, 26–7
 Julian, 20, 22, 27, 118
 Mayan, 137
 Mursi, 20
 Muslim, 23, 26–7
 non-Gregorian, 19
 Old Calendarists, 26
 religious, 54
 secular, 26
 Trobriand, 18–20
Calendrical Calculations, 26–7
calypso, 109–11
canonical hours, 13–14. See also Nones
capitalism, 7, 22, 50–2, 65–8, 129
Caribbean (applies to the region), x, 27, 51–4, 64, 67–8, 71, 88–9, 101, 125. See also West Indies (applies to the Anglophone Caribbean)
Carnival, 63–4, 80, 109–12
Carr, D., 132

INDEX 161

Castells, Manuel, 25
Chakrabarty, Dipesh, 69, 114, 126
Chambers, George, 105
Chance, John, 125
change, 2, 34, 93, 97, 99, 105, 112, 126, 133–8
Cheney, Christopher, 27
Christianity, 13, 25, 27, 30, 94, 95–6, 98–9, 126
chronobiology, xi–xii, 16, 31, 35–45, 68, 90, 93, 99, 137, 140
chronology, 16, 23–5, 72–4, 76, 79–80, 81, 83, 88–91, 93, 99–100, 101, 112, 126–8, 135, 138–9
  Chinese, 26
  Christian, 29–30, 126, 135
  European, 23, 27, 30, 90, 139
  Hindu, 26
  Jewish, 26
  Muslim, 26
  secular, 29
chronometer, 13
chronotope, 73, 96, 124
circadian rhythms and cycles, 11, 30–1, 33, 37–8, 41–5
Circular T, 6 (*see also* International Bureau of Weights and Measures)
Clark, Andy, 22
clepsydrae, xii
Clifford, James, 2
climate change, 96–9, 112, 140
clock, x–xii, 1, 3–11, 14, 16, 18, 21, 29, 30, 31, 33–4, 36–8, 40–5, 47, 49, 58, 99, 101, 115, 118–19, 139–40
  analog, xi
  atomic, 5, 6
  counterclockwise, xi
  mechanical, 3, 13, 36–8, 41
clock time, 3, 4–5, 8, 10, 13, 15, 18–19, 22, 24, 28, 30, 33–4, 38, 40, 42, 44, 47–8, 55, 57, 65, 69, 90, 94–5, 115–16, 140

CNRS, 5
coevalness, 119–39
cognitive extension, 22
cognitive mediation, 22
cognitive tools, 33, 99, 119
Cole, Jennifer, 73, 89, 128
Collins, John, ix, 3
colonialism, xi, 27, 49, 54, 83, 89, 102–3, 116, 118, 128–9, 139
commodification, 4, 51–3, 63–5, 68, 116
commodities, 47, 50, 52, 66, 103–4
computers, 24, 41, 65
Condon, Richard, 43
contemporaneity, 71, 130–1, 137
contingency, x–xi, 15, 69, 129, 136
continuity, x, xii, 8, 94, 99, 100, 115, 125–7
Conway, M. A., 73
Cook, James, 137
cooking, 49, 82–3, 87
Coordinated Universal Time, xiii, 6, 30
Corbin, Alain, 14
counting, xii, 10, 19, 20, 23, 26, 35
Craig, Susan, 63
Crapanzano, Vincent, ix, 93–4, 96, 98, 124
crisis, 2, 72, 97, 100, 102, 115–16
Csatáry, Lázló, 50
cultural choice, 23
cultural conservatism, 98–9
cultural differences, 1, 18, 20, 28, 118, 131
cultural logic, 2, 20, 36, 47–8, 53, 64, 68
cultural relativism, 133
cultural reproduction, 98–9
cultural variability, 2, 18, 23, 74, 89, 123, 129, 131–6
Cuthbert, Saint, 25

cycles, xi, 5, 14–15, 18, 22–3, 29, 36–8, 42, 47–8, 53, 64, 66, 89, 97, 99, 103–4, 105, 110–11, 112, 129, 133, 138, 140
  annual, 3, 18–19, 23
  biological, 35–8, 40
  environmental, 15, 34, 36–8, 45, 116
  free-running, 41–2
  light/dark, 14, 33, 39, 42, 44 (*see also* circadian rhythms and cycles)

D

Daily Telegraph, 20
Dallmann, Robert, 34
date, 19–20, 22–3, 25, 27, 50, 69, 77–81, 83–4, 88, 90, 100, 115, 122, 127, 134–5, 139
Davis, J., 133
day, 5–6, 11, 15, 21, 26, 27, 40, 42, 47, 116 (*see also* length of day)
  Egyptian, xi
  Jewish, 26
  mean solar, 5–6, 49
  rotational, 6, 49, 118
  secular
  solar, 5, 29, 118
daylight, 13–14, 23, 27, 31, 33, 39–42, 51
Declerq, Georges, 29
deconstruction, 4, 15
Deleuze, Gilles, 95, 136
Derrida, Jacques, 4–5, 7–8, 12, 14–16, 48, 137, 140
Dershowitz, Nachum, 26
diaries, 94
Dionysius Exiguus, 29, 134–5
Dorhn van-Rossum, Gerhard, 14
Dreaming, The, 133
duplicity, 5–7, 9, 13–15

duration, 3–6, 20, 23, 26, 29, 32, 34–5, 39, 45, 48–54, 57, 60–1, 63–4, 66–7, 90, 93–4, 118, 122, 134
Durkheim, Emile, 3

E

Earth, xi, 5–6, 49, 97, 115
Earth rotation, 5–6, 30, 116
economic development, 51–3, 63, 68, 103
economics, xi, 45, 50–8, 61–2, 64–8, 94–9, 102–3, 106, 112–13, 116, 137
  classical, 50, 52–3, 56, 58, 60–3, 66–7, 74
  neoclassical, 50, 52–3, 62–3, 67
economists, 47, 61, 66
Edensor, Tim, 90
education, x, 4, 59, 86, 119
Einstein, Albert, 16, 69, 114–15
electromagnetism, 36–7
Elias, Norbert, 4, 8, 34
elision, x, 34, 48–9, 68, 90, 98, 101
empire, 50, 95
endings, 72, 100, 105, 134
Enlightenment, The. *See* European Enlightenment
environment, 22, 35–40, 42–5, 116, 140
episodic, 73, 95–6, 101, 105
epistemology, 22, 99, 121, 123, 126
epoch, 1, 48, 65, 79, 117
equator, 23, 31, 40
eternity/eternal, 133, 135–6, 138
ethnographer, xi, 2, 18, 28, 48, 123–6, 130, 134, 136–7
ethnographic evidence, 43, 65
ethnographic interview, 119
ethnographic present, 2, 17, 28, 123–4, 131

ethnography, ix, x, xii, 2–3, 16, 17–23, 28, 35, 48, 68, 77, 88, 115, 118, 123–30, 132–9
Europe, 13–14, 18, 33, 49–50, 89–90, 113, 116, 126–7, 129
European educational system, 119
European Enlightenment, 94, 113, 140
European philosophy, 131–3, 136
European post-Enlightenment thought, 2, 16, 113, 118
European representations of history, x, 90, 114, 134, 139
European representations of time, x, xi, 1, 3–4, 14, 18–24, 27–30, 48–9, 113, 118–19, 134, 138, 140
Evans-Pritchard, E. E., 19, 89, 131
exchange value, 47, 52, 60, 64

F
Fabian, Johannes, 2, 17, 28, 118–19, 121–8, 130, 133, 135, 137
factories, 7, 58, 66–7, 77, 84, 86–7, 104, 106
Fairfax, Thomas, 11
Fanon, Frantz, 54, 89–90
fashionably late, 12. *See also* lateness
Feeney, Denis, 25–6
Ferry, Elizabeth, 74
field notes, 100, 123
field research, 17, 35, 38, 44, 119
Fielding, Henry, 95–6
fieldwork, 7, 17–18, 100, 106, 119, 124, 128, 130
Fitzgerald, Robert, 48
flexibility, 8, 43–4
food availability, 39, 43
Forte, Maximilian, 89
Foucault, Michel, 116, 125–7
Frazer, James, 18, 22

Freeman, Karen, 11
Freilich, Morris, 67
Freire-Marreco, Barbara, 18
French Revolution Decimal Time, xi, 9
Friedman, Thomas, 25
future, 16, 21, 26, 34, 69, 72, 89–90, 93–116, 117, 119, 124, 128–30, 132
future orientation, 129

G
Galison, Peter, 24, 34
Geertz, Clifford, 124, 131, 136
Gell, Alfred, 22, 134
genealogical method, 18, 20
Glennie, Paul, 47
Glissant, Édouard, x, 93, 101, 125–6
global time administration, 11, 13. *See also* International Bureau of Weights and Measures
globalization, xii, 5, 24–6, 30–1, 49, 67, 83, 95, 104, 112–13, 116
globe, 25, 28, 42–3, 78, 95
Gomes, Alfred, 103
Gompf, Heinrich, 43
Goodman, Jane, 129
Gore, Al, 97
Gospel of John, 24–5
government, 54–5, 62–4, 68, 83–4, 103–8, 110–11, 114
Greenhouse, Carol, x, 2–3, 21, 95
Greenwich, England, 115
Gurevich, A. J., 1, 3
Guyer, Jane, 94–6, 98, 117

H
Habermas, Jürgen, 95
hagiography, 25
Halbwachs, Maurice, 78, 86

Hallowell, A. Irving, ix, 131
Hamilton, Robert, 11
Harris, Wilson, x, 101
Harvey, David, 25
Hassan, Robert, 44
health, 25, 34, 43
Hegel, G. W. F., 136
hegemony, 26, 67, 116
   Christian, 30
   colonial, xi
   European, x–xi
   homochronic, 117–40
Heidegger, Martin, 131–2
Henzi, S. P., 33
Hesiod, 23
heterochronicity. *See also* temporal pluralism
high-frequency trading, 50, 65–6
Hintzen, Percy, 63
historians, 2, 27, 47
historically significant events, 75–84, 86–8, 122, 128. *See also* temporal landmark
historiography, 3, 65, 71–3, 90, 125, 127, 136, 138–9
Hobsbawm, Eric, 72, 89
Hodder, Ian, 125
Hodges, Matt, 115
holidays, 22, 27, 53–4
holism, xi
homochronism/homochronicity, x, 28–9, 30–1, 34, 43–4, 45, 49, 53, 68–9, 71–3, 88, 90, 93, 99, 100–1, 110–16, 123, 126, 128, 129, 131, 135–8, 140
homogeneous, empty time, 15, 68–9, 71–2, 88, 90, 100–1, 110–16
hope, 34, 93–5, 105, 112
horological revolution, 8
horology, xi, 37, 44. *See also* clock
Hoskins, Janet, 89

hour, xi, 6–7, 9, 14–15, 24–5, 27, 34, 47, 50, 53, 58, 65, 66. *See also* canonical hours; zmanim
"Hours don't make work", 56
Howe, Leopold, 21, 136
Hubert, Henri, 1
Hume, Lynne, 133
humidity, 35–6, 140
Hunnicutt, Benjamin, 51
Hutton, Ronald, 22
Huygens, Christian, 118

I
illusion/illusory, 69, 115, 134
imagination, 16, 20, 71–2, 89, 93–6, 98–101, 109, 111–16, 128–9, 132
IMF. *See* International Monetary Fund
immanence, 97–8, 115
imminence, 97–8
International Bureau of Weights and Measures (BIPM), 6
International Monetary Fund (IMF), 62, 102, 106, 111
intersubjectivity, 68, 74, 80, 88–9, 91, 119, 123, 126, 128–31, 134
invention of tradition, 89
irregularity, 5, 15, 29–30, 42, 78, 116
Isidore of Seville, 24–5
Islamic prayer times, 27

J
James, C. L. R., x, 54, 109
James, Jason, 73
James, William, 128
Jameson, Frederic, 135
Jarjisian, Stephen, 43
Jemmat, Catherine, 11
Johnson, Maynard, 36

Johnson, Samuel, 10–11
Journal of Biological Rhythms, 42–3
"jus' now", 60, 74, 121
justice, 49–50
jyotish, 27

K
kairos, 49–50, 53, 63–8, 73, 93, 117.
   See also moments; timing
Kanigel, Robert, 48
Kant, Immanuel, 132
Kellogg Company, 51
Kelly, John, 69
Kermode, Frank, 71–2, 95, 100, 134
Kinneavy, James, 49
kinship, 18, 89
Koepping, Klaus-Peter, x
Kojève, Alexandre, 89
Koselleck, Reinhart, 129
Kuhn, Thomas, 34
Kula, 21
Kurbat, M. A., 73

L
labor, 33, 45, 50–6, 58–9, 61–5,
   67–8, 74, 88–9, 94, 104, 112,
   137
  abstract, 52, 67–8
  agrarian/agricultural, 7, 106
  alienation of, 51, 67
  average labor, 48, 50–1, 56, 58,
   60–3, 66–8
  capitalist, 7, 47–8, 65, 67
  commodified, 52–3, 63–6, 68, 116
  costs, 51–3, 62, 107
  industrial, 47, 67
  relations, 7, 88, 116
  representation of, 47–8, 51–3,
   58–9, 60–2, 65, 93, 99

surplus, 51–3, 58, 62
unlimited supplies of, 51–3 (*see also*
   Lewis, W. Arthur)
wage, 47, 52–3, 56, 59, 62, 106,
   108
labor time, 16, 35, 47–8, 50–3, 56,
   58, 60–1, 66–8, 90
laboratory research, 6, 34, 38–40,
   42–5, 137
Lacan, Jacques, 89
Laing, R. D., x
Lakoff, Andrew, 117
Landes, David, 116
language, 3, 21
lateness, 11–12, 120
latitude, 9, 26, 27, 31, 39, 50–1
Latour, Bruno, 34, 115
law, 10, 50, 79, 94
Lazar, Sian, 65, 93
Leach, Edmund, 20
leap second, xii
leap year, 26, 64
Lederman, Rena, 23
Leliavski, Alexei, 42
length of day, 5–6, 27, 42–3
Levine, Judah, 6
Levine, Robert, 1, 3
Levitas, Ruth, 93
Lewis, H. Gregg, 53
Lewis, W. Arthur, 51–3, 63, 68
life cycle, 21, 77–9, 84–8
life stage, 78–9, 86–8
light, 23, 30–1, 36, 39–44, 49 (*see also*
   daylight)
  artificial, 40, 44
Limbert, Mandana, 132
liming, 77
linear time. *See* time—linear
linguistic space-time mapping, 21
Livingstone, Frank, 37
"long time", 60, 74–8, 83, 120–2

longitude, 31
lottery, 114
Luhmann, Niklas, 117
lunations, 17–20
Lyotard, Jean-François, 135

M
Macey, Samuel, 8
malaria, 35, 86
Malinowski, Bronislaw, 2, 17–20
man hour. *See* person hours
Manoogian, Emily, 43
Marcus, George, 2
Marcuse, Herbert, 95
Marramao, Giacomo, 16, 64
Marx, Karl, 47–8, 52–3, 58–9, 61, 65–8, 127
McIntosh, Ian, 133
McLuhan, Marshall, x, 99
Mead, George Herbert, 128
mean solar time, 5–6, 8, 14, 47, 49
meaning, ix, 4–5, 10, 14, 19, 60, 65, 68, 73, 75, 90, 94, 101, 119, 132, 134, 138
measure/measurement, 5, 23, 38, 44, 48, 50, 52, 56, 58, 64, 67, 68. *See also* time measurement
medication, timing of, 34
memory, ix, 72–4, 77–8, 80–2, 84–9, 91, 128
  clustering of memories, 73–4, 78, 80–1, 84–6, 88–9
Menaker, Michael, 41
metaphor, ix, xi–xii, 3, 21, 36–7, 44, 94
  clock, xii, 36–8, 41–5
  container, 50, 93
  spatial, 3, 21
methodology, 2, 18, 20, 24, 38, 44, 48, 72, 93, 129

mice, 36, 39
middles, 72, 100, 134
midnight, 7, 26
Milham, Willis, 11
Miller, Daniel, 136
Mills, David, 41
Mintz, Sidney, 54, 67, 88, 138–9
minute, 6–7, 31
Miyazaki, Hirokazu, 93–5, 117
modernity, 5, 14–16, 27, 69, 71, 93–4, 101–2, 113
moments, 23, 25, 34, 45, 48–50, 54, 60, 63, 65, 68, 72–4, 79, 81, 97, 100–1, 117, 121
month, 15, 18–20, 22–3
Monthly Magainze and British Register, The, 11
Moon, 6, 18–20
morality, 12, 14, 58, 79, 112, 121–2, 133, 138
mosquitos, 35–7, 45
Mulla, Sameena, 115
multiple times. *See* temporal pluralism
Mumford, Lewis, 99
Munn, Nancy, 21, 131, 134
Murakami, Kyoko, 73
myth, 72, 89–90, 138

N
nanosecond, 65
NAR. *See* National Alliance for Reconstruction
narrative, 65, 71–5, 78, 84, 88–90, 94–6, 100–16, 122–5, 127, 132, 134, 138–9
National Alliance for Reconstruction, 105–6, 111
nationalism, 68–9, 71, 113
natural, x, 15, 30, 47, 65, 69, 97, 114–15, 118, 137

INDEX 167

naturalized, x, 4, 16, 18, 23, 69, 118, 140
nature, 3, 29–31, 34, 44, 69, 114
Nelson, R. A., 5
New York Times, 11
Newcomb, Simon, 5
Newton, Isaac, 15, 34, 39, 116, 118, 132
Nielsen, Morten, 21
Nilsson, Martin, 22, 23, 29, 35–6
Nones, 13–14
noon, 7, 8–15
"noon at two o'clock", 4–15
noon mark, 7, 9
Northcott, C. H., 59, 61
Notes and Queries, 18
Nowotny, Helga, 25, 38
NTP (Network Time Protocol), 41, 43
Nugent, David, 94
numbered versus measured, 64

O

Obeyesekere, Gananath, 137
Oil Boom, 74, 82, 84, 86, 88, 102, 104–6, 110
old heads, 75–6, 79, 86–8
ontology, ix, 3, 22, 30, 69, 114, 117, 131–3, 135, 137–8
Opitz, Sven, 94

P

Panday, Basdeo, 106
Paneth, N., 37
Passion of Jesus, 14
past, 1, 21, 34, 61, 68, 71–91, 93–7, 99–105, 107–8, 110–12, 117, 119–22, 124–6, 128–30, 132–4, 137–9

patronage, 63–4, 68, 81
Peel, J. D. Y., 136
People's National Movement (PNM), 102–6
perception, ix, xi, 94
person hours, 50
phase relationship, 33, 41, 43, 48–9
phenomenological/phenomenology, 69, 113, 121–3, 128–31, 137
physics, xii, 8, 16. *See also* Einstein, Albert; Newton, Isaac; relativity
Piketty, Thomas, 68
Pittendrigh, Colin S., 35–8, 140
planning, 11, 72, 93, 95, 98, 100, 102, 107–9, 111–12, 129, 140
plantations, 7–8, 67–8, 83, 126–7
Plato, 49, 134–6
Pliny the Elder, 27
Plummer, Denyse, 110
PNM. *See* People's National Movement
Poe, Edgar Allan, 9–10, 12, 73
polysemy, 5, 117
Pope, Alexander, 10–13
postcolonialism, x, 90, 102–3, 111–13, 128
Postill, John, 3
Postone, Moishe, 67
Powdermaker, Hortense, 17
power, 30, 61, 67, 69, 95, 111, 137–8
precapitalist societies, 65, 129
precision, 6, 8, 23, 44, 65
predictability, 73–4, 97, 99–100, 104, 111–12. *See also* unpredictability
prefiguration, 101, 105, 110, 111, 113–16
pregnancy, 34
present, 69, 72–3, 75–7, 79–80, 91, 93–103, 105, 107–12, 115, 117, 119, 120–6, 132–4, 137–8
phenomenological, 123, 128–31

Price, Richard, 71, 139
primatology, 33, 38
privileged ideas of time, 18, 21, 45, 49, 50–1, 66, 68, 90, 94, 114–15, 118–19, 134–7
productivity, 51, 53, 56, 58, 60–1, 65, 106, 108
psychology, 12, 36, 41, 68, 73–4, 29
psychophysics, 1, 90, 99
Ptolemy, xi
public policy, xii, 97–9, 102–3, 105–6, 109, 113, 116, 140

R
Rabinow, Paul, 124
Rackham, H., 27
Rappaport, Joanne, 73, 127, 137
rationality, x, 65, 115, 138
real time, 3, 72, 95
recession, 99–102
reductionism, xii
relativity, 4, 7, 16, 69. *See also* Einstein, Albert
rhetoric, 49, 66, 79, 83, 90, 123–4, 130
Rhys, John, 22
rhythms, x–xi, 11, 15, 28, 30, 34–7, 40, 42–4, 53, 59, 63, 68, 102, 126, 140
Rice, Prudence, 137
Ricoeur, Paul, 71, 132, 135
Ringel, Felix, 93–4
Rivers, W. H. R., 18, 20
Robbins, Joel, 117
Robinson, A. N. R., 105–6, 111
Robinson, John, 73
Robinson, John A. T., 65
Rodney, Walter, 54
Roediger, David, 116
Roman Empire, 50

Rowntree, Joseph, 48
Rowntree, Seebohm, 48, 59
Rowntree Company, 48, 59, 61. *See also* Northcott, C. H.
Ryan, Selwyn, 63

S
Sahlins, Marshall, 137
Samhain, 22
Sanjek, Roger, 17, 123
Sarte, Jean-Paul, 89
schedule, 12, 45, 50, 54–5, 59, 99
Schutz, Alfred, 130, 137
science, ix, xii, 5–6, 15, 30, 34, 42, 95, 113, 115. *See also* physics; psychology; psychophysics; time metrology
scientific management, 47–8, 58–9. *See also* Cadbury, Edward; Northcott, C. H.; Rowntree Company; Taylor, Frederick Winslow
Scipio, 50–1
SCN. *See* suprachiasmatic nucleus
Scott, David, 71, 90
Seaman, Robert, 6
seasonal affective disorder, 43
seasonal variations, 31, 33–4, 42–3, 49, 51, 65
seasons, 21, 42
second, 5–6, 24, 31, 65
  definition of, 5–6
  International System (SI), 5
secularism, 26, 71, 94, 98, 114, 126, 138
Sellix, Michael, 43
sequence/sequential, x, 2, 24, 80, 90, 96, 117, 126, 132, 134, 140
Shaw, Matthew, 9
Sherman, Stuart, 94

INDEX 169

Shum, Michael, 73
signification, 4, 5, 8, 13, 16, 24, 69,
    105, 115, 117, 140
simultaneity, 4, 24–5, 69, 88, 90,
    115, 117
Singer, S. Fred, 97
Sinha, Chris, 3, 20–1
Smith, Adam, 52
Smith, Mark M., 116
social science/social theory, ix, xi, 1, 3,
    7, 16, 71, 131, 133, 135
solar time, 31
Spectator, The, 12
speech acts, 24
spontaneity, 54
stability, 29, 94
standardization/standards, 5, 26,
    30, 41, 48, 50, 53, 95, 108,
    116, 140
Stanner, W. E. H., 133
stasis, 135–6, 138
Stern, Pamela, 42
Stern, Sacha, 1, 3, 19
Stewart, Charles, 72, 95
Stiegler, Bernard, ix
stock trading. *See also* high-frequency
    trading
sundial, 9, 11
sundown, 26, 35
Superblue, 110
suprachiasmatic nucleus (SCN), 41–5
surplus value, 52–3, 58, 62
synchronization, 9, 14, 24, 41, 43

T
Taft, Robert, 14
Taussig, Michael, 127
Taylor, Charles, 15, 71, 101
Taylor, Frederick Winslow, 47–8, 59
technology, 3, 5–6, 23, 97–9

technoscience, 95
teleology, 110–11, 135
temperature, 23, 36–7, 39
temporal anchors, 29, 83–4, 90
temporal assumptions, x–xii, 1–4,
    14, 16, 20–2, 28–30, 50, 68,
    71, 74, 89–90, 97, 114, 117,
    129, 136
temporal conflict, 101–2, 132
temporal framework, 23, 25, 90, 121
temporal grid, 23–5
temporal lacunae, 74, 89–90, 96, 101
temporal landmark, 73–4, 77–8, 80,
    86, 88–91, 101
temporal niche, 33, 35, 37, 43
temporal orientation, 58, 61, 79, 94,
    122
temporal pluralism, xi, 7, 10, 21, 55,
    96, 98–9, 100–2, 115, 123,
    125–9, 137–9
temporal regularity, 1, 74, 96
temporal uniformity, x, 4, 8, 10,
    13, 15, 23, 25, 27, 34, 39,
    42, 44–5, 48, 50–3, 64,
    68–9, 73–4, 88–91, 93–6, 99,
    109, 112, 114–16, 118, 126,
    129, 140
temporal universals, 3, 128, 132
temporal warping, 2, 33, 73
temporality, xi, 2–3, 16, 22, 27, 30,
    40, 65, 68–9, 71, 79, 93–7,
    100–3, 107, 109–16, 117–20,
    123–5, 129, 131, 135, 137,
    139–40
    anthropology, 2–3, 123–5, 136–7
    European, x, 94, 118–19
    narrative, 65, 71–2, 89, 100–5, 111
    postcolonial, x, 111
    secular, 94
Thanksgiving, 49
Thompson, E. P., 7, 47, 65, 67, 116

Thucydides, 25–6
Tignor, Robert, 68
Tillich, Paul, 65
Time (*see also* calendar time; chronology; clock time; homochronicity; labor time)
  circular, 21
  definition of, 5–7, 25, 55, 94
  division of, xi, 1, 27, 90
  linear, 21
  "no concept of", 20–1
  uniform (*see* temporal uniformity)
time discipline, 30, 51, 59, 120
time maps, 73
time measurement, 5–7, 15, 18, 23, 30, 33–4, 38, 44, 47–8, 53, 56–8, 64, 67, 69, 116, 118
time metrology, xii, 44, 95
time reckoning, 9, 13, 18–19, 22–3, 29, 57, 88, 116, 137, 140
  European, xi, 21, 23, 27–30
time standard, 5, 26, 30, 41, 95, 116, 140. *See also* time metrology
time zone, 9–10, 15, 31, 116
time–space compression, 25
timelessness, 17, 28
timescale, 5, 10, 29, 30, 88, 140
timestamp, 65
timing, 33–4, 48–51, 54, 59–62, 64–5, 67–8, 75, 86, 110 (*see also* kairos)
  biological, 37–8, 41–5
  psychological, xii, 36, 41
Traweek, Sharon, 34
Trinidad, x, 7–8, 28, 35–6, 45, 50–1, 53–5, 58–68, 73–89, 99–116, 119–22, 126, 136
Trouillot, Michel-Rolph, 126, 130
Turton, David, 20
twilight, 26

U
ULF. *See* United Labour Front
uniformitarianism, 97
uniformity, x, 8, 27, 51–2, 109. *See also* temporal uniformity
United Labour Front, 106
United States of America, 49, 66–7, 84, 103, 106
unpredictability, 103–5, 110–12, 140
use value, 52, 60, 62, 64
U.S. Department of Defense, 98, 112
UTC. *See* Coordinated Universal Time
utopia, 93

V
Valéry, Paul, 117
Vegetius, 50
Vélez-Ibáñez, Carlos, 74
Verne, Jules, 12–13, 25, 73
Vico, Giambattista, 24
Virilio, Paul, 66, 95

W
Wackermann, Jiří, xi–xii
wages, 47, 52–3, 56, 59, 62–3, 82, 86, 107
Warman, G. R., 37
watches, 6–12, 14, 17–18, 24, 30, 36–8, 57
Watchman, 110
West Indies (refers to Anglophone Caribbean), x, 53–4, 64, 68, 89
West-Pavlov, Russell, 71–2
Western assumptions, x, 21, 28, 65, 72, 125, 135
Wever, Rütger, 37
Whe Whe, 113–15
White, Geoffrey, 73
White, Hayden, 71

Wickman, Matthew, 69
Wilcox, Donald, 24, 90
Williams, Eric, 102–5, 107, 109, 111
Wonderland, 28
workday, 51, 56
 six hour, 51
World Bank, 62
World War II, 50, 73, 76, 80–2, 84, 86–7
Wynter, Sylvia, x, 117–18

Y
year, 5, 15, 20, 64, 68, 76, 80
 lunar, 18–20, 23
 solar, 19–20, 23, 29
"years aback", 75–9, 112, 121–2

Z
Zerubavel, Eviatar, 9, 34, 73
zmanim, 27

The manufacturer's authorised representative in the EU is Springer Nature Customer Service Centre GmbH, Europaplatz 3, 69115 Heidelberg, Germany. If you have any concerns regarding our products, please contact ProductSafety@springernature.com

Printed and bound by CPI Group (UK) Ltd, Croydon, CR0 4YY
26/03/2026
02078762-0001